1 MONTH OF FREE READING

at

www.ForgottenBooks.com

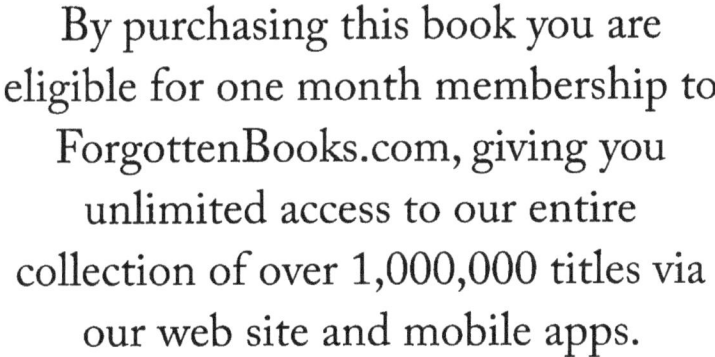

By purchasing this book you are eligible for one month membership to ForgottenBooks.com, giving you unlimited access to our entire collection of over 1,000,000 titles via our web site and mobile apps.

To claim your free month visit: www.forgottenbooks.com/free148813

ISBN 978-1-5284-8506-7
PIBN 10148813

THE DIAGNOSIS OF DISEASES OF THE CORD.

LOCATION OF LESIONS.

BY

DR. GRASSET,

CLINICAL PROFESSOR AT THE UNIVERSITY OF MONTPELLIER, ASSOCIATE
OF THE NATIONAL ACADEMY OF MEDICINE,
LAUREATE OF THE INSTITUTE.

TRANSLATED BY

JEANNE C. SOLIS, M.D.,

DEMONSTRATOR OF NERVOUS DISEASES AND ELECTRO-THERAPEUTICS
IN THE UNIVERSITY OF MICHIGAN.

GEORGE WAHR,
PUBLISHER,
ANN ARBOR, MICHIGAN.

THE DIAGNOSIS

OF

DISEASES OF THE CORD.

INTRODUCTION.

Given a patient in whom a disease of the cord has been recognized, how can the exact location of the medullary change be determined clinically?

What system or systems of the cord are exclusively or principally attacked?

At what level of the spinal axis is the lesion located?

I wish to sum up here the elements of the answer the present neuro-pathology permits to this ques-. tion, which is an interesting one to practitioners generally.

For if this chapter of clinical geography of the cord, founded by the chiefs of the French neuro-pathological school, Duchenne, of Boulogne, Vulpian and Charcot, seemed at the beginning a chapter of pure science, today it has been so enlarged, confirmed and made so exact that it is now absolutely practical, accessible and useful to all.

In the first place the necessary indications for surgical intervention were found here. This field increases every day in proportion as the operations become less dangerous and their technique is perfected.

Further, the different conditions called classically diseases of the cord are anatomico-clinical syndromes characterized by the fixity of their symp-

toms and the fixity of the location of the corresponding lesion.   Whence it results that the diagnosis of
this location of the lesion constitutes the complete
diagnosis of this syndrome.

Thus to recognize progressive muscular atrophy
or tabes, it is sufficient to recognize that the lesion,
in the patient examined, is located in the anterior
horns of the grey substance or in the posterior bundles.

Then, without denying the importance of anatomical and nosological diagnoses which when possible should complete the physiological, it may be
said that the physiological diagnosis of the location
of the lesion is absolutely of primary necessity for
all physicians today.

The natural division of this little book is into
two chapters.   1st. In the first chapter we will study
the semeiology of the systems of the cord, that is to
say, the signs by which is recognized the location
of the medullary change in such or such system of
this organ (anterior horns, posterior horns, posterior columns, etc).   2nd. In the second chapter we
will seek to make a diagnosis of the location of the
lesions.

The study of the clinical anatomy of the cord
ought to be the appointed prelude and indispensable
basis for pathology.   For it is useless to refer to the
ordinary anatomical description. *Anatomical anatomy*
is useful to the clinician; it is the foundation.   But
physiological anatomy is still more necessary.   A
symptom is a function pathologically deviated.   Then
it is necessary to have a basis of functional or physiological groupings of organs to make a work useful in
practical medicine.

This study which requires too much to be made
here, will be the object of a special publication.

I. THE DIAGNOSIS OF DISEASE OF THE MEDULLARY SYSTEM.

We will study successively in this chapter the eight following syndromes:

1. The syndrome of the posterior columns: sensory troubles and ataxia;

2. The syndrome of the antero-lateral columns: pareto-spasmodic state, contractures and intentional tremor.

3. The associated syndrome of the posterior and lateral columns: ataxo-spasmod.c state;

4. The syndrome of the anterior horns: muscular atrophy;

5. The associated syndrome of the anterior horns and the lateral columns: spastic muscular atrophy;

6. The syndrome of the centro-posterior grey substance: dissociation of sensation, called syringo-myelic (and vaso-motor troubles);

7. The associated syndrome of the anterior horns' and of the centro-posterior grey substance (syndrome of the whole grey substance): muscular atrophy, dissociation of sensation, called syringo-myelic, and vaso-motor troubles;

8. The syndrome of the lateral half of the cord: hemiparaplegia crossed.

For each syndrome we will study successively: 1st. The group of cases in which the lesion is limited to this system (lesions and symptoms); 2nd. Those in which the lesion attacks this system without being exclusively limited to it (lesions and symptoms); 3rd. The synthesis of the syndrome (clinical description, pathological physiology and differential diagnosis).

1. The Syndrome of the Posterior Columns: Sensory Troubles and Ataxia.

A. There is only one group of cases in which the lesion is systematically *limited* to the *posterior columns :tabes or progressive locomotor ataxia.*

Let us sum up the *lesions* and *symptoms*.

1. Although a gelatinous degeneration of the posterior columns of the cord was anatomically noted by Hutin in 1827, it may be said that the pathological anatomy of tabes began with Bourdon and Luys (1860) a short time after the masterly clinical description of Duchenne (1858).

In the first period a primary systematic sclerosis of the posterior columns in their entirety was claimed; in the second period (Charcot and Pierret, 1871) more was determined: The principal, initial lesson was localized in the external part of the posterior columns (posterior root zones).

Finally in the third period, the lesion of the posterior roots appeared most constant (Leyden and Vulpian) ; then the starting point of the lesion was placed in the ganglions (P. Marie, 1892) and·tabes was made a disease of the sensory protoneurone (Brissaud 1895, deMassary[1] 1896).

In fact, the lesions of tabes, in the beginning is localized in the external bundle of Charcot and Pierret; then in the more advanced stage it includes the zone of the entrance of the posterior roots of Philippe, that is the zone of Lissauer and the cornuroot zone of Marie. Finally in a case of long dura-

---

[1]MASSARY (DE). Le tabes dors. dégenér. du protoneur. centrip. Th. Paris, 1896.—Voir aussi pour ce paragraphe: PHILIPPE. Le tabes dorsalis. Paris, 1897: et GEREST. Les affections nerveuses systématiques et la théorie des neurones. Paris, 1898.

tion the column of Goll is invaded, especially in the upper portion of the cord. The principal and constant lesion is in the exogenous fibres, root fibres, cylinder axis prolongations of the ganglions. The endogenous fibres often found intact (Marie, Strumpell), have also been found involved (Philippe); the descending fibres at first (triangle of Gombault and Philippe, oval center of Flechsig, postero-internal band, comma tract of Schultze), the ascending fibres later (the cornucommisural zone dying last of the posterior columns). But all these endogenous fibres will be attacked only in the second stage consecutive to the lesion of the exogenous root fibres.

In the grey substance the lesion of the cells of the vesicular column of Clarke is doubtful or inconstant; the alteration of the nerve fibres of this same region (the collaterals given off by the posterior root fibres) is on the contrary very frequent.

The lesion of the posterior root is very frequent but not constant. In the ganglions the cells as a rule are intact.[2]

Let us omit all extra-medullary lesions (nerves, bulb), which are of no interest here.

Brissaud's conception then that tabes is a lesion of the sensory protoneurone may be admitted. Only, if we do not wish to admit a dynamic or unrecognized lesion of the ganglion in many cases, it is necessary to say that the primary, essential location of the lesion in tabes is the intra-medullary part of the sensory protoneurone, that part of the posterior column which we know contains the cylinder axis prolongations of the spinal ganglions.

If with Brissaud we compare the neurone to a tree, the sensory protoneurone is attacked in its

---

[2]Voir sur ce point important: GEREST. *Loc. cit.*, p. 235.

branches. We shall see that in spasmodic tabes the same way the disease in the pyramidal bundle is in the intra-medullary portion of the cylinder axis prolongations of the cortical motor protoneurone. Both ataxic tabes and the spasmodic are diseases of the intra-medullary prolongations of the extra-medullary neurone. (The spinal ganglion for posterior tabes, the cerebral cortex for lateral tabes).

Anatomically this change in the posterior root fibres in tabes involves the myeline especially, and degenerations at first descending and then ascending in the endogenous fibres follow this primary process.

The topography alone of the lesions is what interests us here, as we shall write a chapter on medullary geography alone.

2. We ought now to put together from this location of the lesion the important and essential symptoms of tabes, at least those which are clearly of medullary origin. They may be said to have been nearly all described by Duchenne[3] in his historical memoir of 1858 and by Charcot[4] in his "Lecons de la Salpêtrière."

They may be grouped under the following heads: lightning-like pains, visceral crises (gastric, etc.), girdle sensation, anæsthesias and paræsthesias, plantar anæsthesia in patches, tingling in the forearm, retarded and false localization, abnormal painful persistence of the sensations excited, dissociation (persistence of thermal sensibility), diminution or abolition of the muscular sense, abolition of the patellar tendon reflex (Westphal), of the Achilles

[3]DUCHENNE. De l'ataxie locom. progr. *Arch. gén de méd.* 1858-1859.

[4]CHARCOT. Œuv. compl., t. I, Leç. II, III et IV; t. II. Leç. II et IV.

tendon reflex (Babinski),[5] diminution or abolition of muscular tonicity, hypotonicity (Frenkel),[6] involuntary movements in repose (ataxia of tonicity),[7] paresis or paralysis of the sphincters (vesical troubles), motor incoordination, influence of closing the eyes on the upright position and on the gait (Romberg), trophic troubles, arthropathies, osteopathies, ecchymoses, perforating ulcer.

B. In a series of diseases we find the posterior columns involved, the lesion not being exclusively limited to this system. We will rapidly review them: General paralysis, disseminated sclerosis, Friedreich's disease, syringo-myelia, spinal meningitis.

1. The relation between tabes and general paralysis has been much discussed since Baillarger,[8] that is to say the question of posterior spinal lesions in general paralysis.[9]

First it must be admitted that tabes and general paralysis are two distinct diseases in order that the discussion take place.

In one group of cases, tabes and general paralysis are frequently, according to some (Ballet and Renaud) rarely, according to others (Joffroy and Rabaud) superimposed in the same subject. These cases from this point of view belong to our preced-

---

[5]BABINSKI. *Soc. méd. des. hôp.*, 21 oct. 1898.

[6]FRENKEL. Ueb. Muskel-Schlaffheit (Hypotonie) b. d. tabes dors. *Neurol. Centralbl.* 1896, t. XV, p. 355.

[7]Voir nos leç. sur les mouvem. involont. au repos, chez les tabet. Ataxie du tonus, in Leç. de Clin. méd. 2e série, 1896, p. 271.

[8]BAILLARGER. De la paral. génér. dans ses rapp. avec l'at. locom. *Ann. méd. psychol.* 1862, t. VII.

[9]Voir pour tout ce paragraphe: RABAUD. Contr. à l'ét. des lésions. spin. postér. dans. la paral génér. Th. Paris, 1898.

ing paragraph "A." For tabes is always the same symptomatically and anatomically, whether associated or not with another disease, as general paralysis.

Then there is another group (which is especially interesting here) in which general paralysis (a single disease) has symptoms and lesions of a posterior medullary location.

The lesions in this case closely resemble those of tabes, in a section of cord examined by an expert observer. They differ only in the discontinuity, diffusion and irregularity of the sclerosed zones, the relative or absolute integrity of the posterior roots, and Lissauer's zones, the frequency or constancy of cellular lesions of the grey substance. (Rebaud).

For the symptoms, we find also from the clinical tables such as for a long time have been diagnostic of tabes. But the appearance of cerebral symptoms disturbs the picture. Thus the tendon reflexes then become exaggerated and the special motor troubles of the general paralytic replace the true ataxia with Romberg's symptom.

2. Disseminated sclerosis has for a long time been considered as a disease principally if not exclusively motor because of Charcot's[10] masterly description. But the patches of sclerosis may be located also in the posterior columns; and may simulate to a certain point that of locomotor ataxia; Romberg's sign, motor incoordination, lightning pains, hypæsthesias, and even urinary troubles (Erb Oppenheim).

The same observation, adds Raymond,[11] "was reported to me by one of our present internes of a patient who died during the service of M. Gaucher

[10]CHARCOT. Œuv. compl., t. V, Leç. VI, VII et VIII.
[11]RAYMOND. Leç. sur les mal. du syst. nerv. 1897, t. II, p. 550.

at the hospital of Saint Antoine. During his life this patient had presented, independently of a general spasmodic stiffness, severe pains in the extremities, both upper and lower, imputable to patches of sclerosis in the posterior columns and the corresponding roots."

3. According to the latest works the principal lesion in Friedreich's disease (Hereditary Ataxia) is in the posterior columns, but especially in the column of Goll and also in the column of Gower and the direct cerebellar tracts and their origin in the grey substance (the cells of Clarke's column).

Among the symptoms corresponding to this lesion note: on one hand, an ataxia which is related rather to the cerebellar ataxia than to that of tabes in that it is accompanied by staggering and is only slightly modified by closure of the eyes (the tabeto-cerebellar gait of Charcot), spontaneous movements (ataxia of tonus) and abolition of the tendon reflexes, on the other hand there is usually absence of sensory affections (anaesthesias) and of lightning pains.

4. In syringo-myelia the lesion does not generally affect the posterior columns. But in certain cases these columns may participate in the lesions and these cases pertain here.

Raymond[12] has collected quite a large number of observations on syringo-myelia in which the anaes-thesia was complete instead of dissociated. In these cases the lesion involves the posterior columns.

Such are the cases of Joffroy and Achard, Homen, Oppenheim, and Schuppel.

In a word, Schlesinger, who has made the best

[12]RAYMOND. Leç. sur les mal. du syst. nerv. 1897, 2e série, p. 510.

study of the participation of the posterior columns
in syringo-myelia has shown that in these columns
three regions were especially invaded by the gliomatosis *i. e.* 1st. The part contiguous to the posterior
grey commissure; 2nd. The portions of the column
of Goll adjacent to the posterior median fissure;
3rd. The zone between the columns of Goll and
Burdach.

5. Chronic spinal meningitis and more especially
leptomeningitis (the inflammation of the pia-mater)
are associated with lesions of the posterior columns.
This is the condition in the cases of tabes with con-
comitant meningitis (Vulpian, Dejerine).

6. In ergotism Tuczek[13] described (A) clinically,
"paraesthesias, such as tinglings, numbness, lightning pains, girdle pains, diminution of sensation to
pain, lack of equilibrium when the eyes are closed,"
ataxia and finally abolition of the knee jerks; (B)
anatomically, lesions of Burdach's columns, Goll's
being intact. In a word, symptomatically ·and anatomically tabes.[14]

7. Equally well Tuczek studied the medullary
lesions of pellagra:[15] the posterior columns are attacked in the column of Goll with integrity of the
column of Burdach, the symptoms depending upon
this lesion are nearly none: knee jerk is more often
exaggerated than abolished, no anaesthesia, no true
ataxia except at times in the upper extremities.

In the same part of the posterior cord medullary

[13]Voir P. MARIE in Traité de méd. de Charcot Bouchard,
1894, t. VI, p. 314.
[14]"Rapprocher notre note sur les "Dangers du seigle er-
goté dans l'ataxie locom. progress." *Progrès médical*, 17
mars 1884.
[15]Voir P. MARIE in Traité de méd. 1894, t. VI, p. 319.

lesions have been described in lepra.[16]   But these lesions seem secondary and it is impossible to attribute to them a special symptomatology in the midst of the clinical picture of this disease.

C. With the aid of these various proofs we can now make the synthesis and the pathological physiology of the syndrome of the posterior columns.

The symptoms given may be grouped under two heads: Sensory troubles and ataxia.

The sensory troubles are: Lightning-like pains, paraesthesias, anaesthesias, (especially of the muscular sense) and abolition of the tendon reflexes.

All these symptoms are explained by a lesion of the root fibres of the posterior columns (intramedullary prolongations of the ganglionic sensory protoneurone) or of the first neurones of the relays (ascending prolongations of the neurones of the posterior horns).

The mechanism of ataxia from a lesion of the posterior column has been and still is much discussed.

Walking, all movements, even the most simple in appearance, and the maintaining of the body in whatever position, are really complex acts. Immobility itself is active. It is the cord itself which presides over the coordination of the muscular contractions and relaxations necessary to obtain and maintain each position. Tonus is a part of this general function; it is concerned in the maintenance of immobility in one position and one attitude.

This medullary influence is a reflex one. The centripetal excitation of this reflex comes from the skin, the joints, and especially the muscles. This excitation penetrates the cord by the posterior roots

---

[16]Voir JEANSELME et MARIE. Sur les lés. des cord. postér. dans la moelle des lépreux. *Revue neurol.* 1898, p. 751.

and the root fibres of the posterior columns. It can be comprehended easily how an alteration of these columns seriously interferes with this reflex; whence we have general hypotonia, sphincter troubles, loss of tendon reflexes, and ataxia (of movement and of tonus).

Some centripetal impressions useful in this medullary function of regulation come also by the senses and particularly by sight.

In the normal state these sensorial excitations are secondary and accessory and can at a given moment fail without much interference with equilibrium and the coordination of movements. When, on the contrary, in consequence of a lesion of the posterior columns the excitations normally the chief ones (muscle) fail, the sensorial excitations become important; the patients then use their eyes as crutches (Althaus), and their sudden closure causes an increased disturbance in coordination, Romberg's sign. That is not saying that the ataxic watches or is obliged to watch his feet, but he uses his eyes to take from around him marks and fulcrums to make up for the reflex automatism of his cord. He walks with his brain in place of with his cord. And when this help fails he loses his equilibrium by a kind of sudden vertigo which is Romberg's sign.

From this it is seen that contrary to the classical opinion it is not necessary to consider Romberg's sign as the consequence of the loss or diminution of the muscular sense. I believe I have demonstrated[17] that there is no parallelism nor necessary

[17]Du vert. des atax. (signe de Romberg) in Leç. de clin. méd. 1896, 2e série, p. 312.—Rapprocher cette définition du vertige: "la conscience du trouble de l'équilibre du corps". (FRANK K. HALLOCK, Journ. of nerv. and ment. diseases 1898, p. 175, Anal. in Gaz. hebdom. 1899, p. 82.)

responsibility between Romberg's sign and the state of the muscular sense and that in certain cases of tabes it can be clearly shown that Romberg's sign is present when it is impossible to make out the least diminution of muscular sense by the most delicate tests.

The excitations - from the muscles (as all the others), once in the cord divide: some provoke medullary reflexes, others go to the higher centers where they produce impressions of muscular sense. The first may be alone involved in the lesion of tabes and then incoordination and Romberg's sign are found, but in the same patient the second may ascend to the brain and the muscular sense persist.

It is in this way that cutaneous reflexes can be abolished when the tactile sensibility remains intact.

In other words, in the intra-medullary conduction the reflex paths may be suppressed without the direct paths to the brain being interrupted and then ataxia and Romberg's sign are present without necessarily a disappearance of the muscular sense.

This is so true that the tabetic deprived of his medullary automatic gait continues to walk by his brain. And as in the cord certain fibres can take the place of others[18] the tabetic can rëeducate his cord with his brain.

This explains the success of Frenkel's method in tabes.[19]

---

[18]Dans mon Rapport au Congrès de Moscou, j'ai cité un fait de Erb et de Schultze dans lequel l'ataxie a guéri sans que la lésion des cordons postérieurs ait guéri; il y avait donc eu suppléance, formation de nouvelles voies physiologiques dans la moelle.

[19]Voir le Rapport, cité ci-dessus, sur le traitement du tabes in Leç. de clin. méd. 1898, 3e série, p. 634.

Then, and in conclusion, incoordination and Romberg's sign prove only a change in the posterior root tracts which go to the medullary reflex centers of coordination and tonus. The muscular sense is affected in its turn when the lesion, more extensive, involves also the posterior sensory tracts which go to the higher cerebral centers.

D. For each syndrome studied by us the differential diagnosis consists in indicating the symptoms by which the medullary origin of this syndrome is recognized. That is to say, the symptoms which differentiate it from more or less analogous syndromes of cerebral (or rather intracranial), peripheral (neuritic) or neurosic origin.

1. Lesions of the cerebral cortex may cause sensory troubles and a kind of ataxia, which here is really related to the loss of the muscular sense.[20] This syndrome differs from that of the posterior columns in being strictly hemiplegic, accompanied by other plainly cerebral symptoms (as the "stroke" and hemiplegia) and not accompanied by symptoms plainly medullary spinal (as lightning-like pains, sphincter troubles and abolition of tendon reflexes).

By the same class of differential symptoms may be distinguished lesions of the opto-striate bodies and the internal capsule which also can produce by anaesthesia and post-hemiplegic chorea, a form of ataxia.[21]

The cerebellar syndrome resembles the posterior cord syndrome in many points. But the gait of the little brain is intoxicated, a zig-zag, reeling, staggering one with only a little or no Romberg's sign, no

---

[20]Voir Anesth. d'orig. cortic in *Revue de méd. et de chir.* 1880, n⁰ 2; à la suite du travau de Tripier.

[21]Voir notre travail sur une variété non décrite de phénom. posthémipl. (forme hémiatax.) in *Progrès méd.,* 13 nov. 1880.

lightning-like pains, no sphincter troubles; and on the contrary, from the cephalic lesion there are vomiting and other symptoms from the intracranial vicinity.

2. Peripheral lesions can produce pains (more or less lightning-like) anaesthesia, abolition of tendon reflexes, but not ataxia with the Romberg sign, and no sphincter troubles.

3. For the neuroses, chorea can be easily distinguished since the abnormal movements take place in repose and have a wide range (movements in repose being rare and not extensive in a lesion of the posterior columns).

Hysteria is more difficult of recognition because it can simulate tabes and more often still is associated with it. The distribution of the anaesthesias, the sphincter troubles and the various symptoms of the neurosis (stigmata, attacks) generally permit the diagnosis.

2. The Syndrome of the Antero-Lateral Columns: Paretospasmódic State, Contractures and Intention Tremor.[22]

A. The constituent clinical elements of this syndrome are:

1. Contractures, permanent, variable (sleep, repose, chloroform, Esmarch bandage) or latent (revealing themselves in voluntary movements): it is the type of the manifestation of a lesion or the absence of the pyramidal bundles.

[22]Voir nos leç. sur les contractures et la portion spinale du faisceau pyram. (le syndr. parétospasm. et le cordon latéral), in *Nouveau Montpellier méd.* 1899, janvier à mars; et une Note sur les contractures et la portion spinale du faisceau pyram. in *Revue neurol.* 1899, p. 122.

2. With the least intensity and less limitation of the lesion there is paresis with exaggeration of the tendon reflexes (patellar tendon reflex, tendon Achilles studied by Babinski), clonic phenomena (clonus or epileptoid tremor of the foot, knee clonus) the phenomena of the toes (Babinski), extension of the toes on excitation of the sole of the foot.[23]

3. When the lesion of the antero-lateral columns leaves some fibres intact in the midst of the sclerosis, an intentional tremor, typical of disseminated sclerosis, absent in repose, develops on action and is increased by a repetition of the act.

B. This system is exclusively attacked and in consequence the syndrome is pure in three diseases: late contractures of hemiplegics, spasmodic tabes, (ataxic paraplegia), and Little's disease.

1. Described by Suavages among paralytic contractures, late contracture of hemiplegics is separated from the early contractures by Todd (1856) and joined by Charcot and Bouchard (1866) to the descending degeneration of the pyramidal bundle already described by Cruveilhier above the pyramids and by Turck (1851) below. Next comes Brissaud's masterly study (1880).

It may be said that for all neurologists (we shall discuss in a paragraph of the pathological physiology the opposed opinion of van Gehuchten) the late permanent contracture of hemiplegics with ex-

---

[23]Nous rapprochons tous ces éléments dans le même groupe symptomatique, malgré les publications de Maurice de Fleury (1884) et de van Gehucten (1897).—Pour tous ces phénomènes, voir: STERNBERG. Die Sehnenrefl. u ihre Bedeut f. d. Pathol. d. Nervensystems, Leipzig, 1893; et GARNAULT. Contrib. à l'ét. de quelques réfl. dans l'hémipl. de cause organ. Th. Paris 1898.

aggeration of the tendon reflexes is the syndrome of a lesion of the pyramidal fibres, consecutive to one of a cerebral center.

2. Spasmodic tabes has been more discussed.

Erb and Charcot described the syndrome in 1875 under the name of spastic spinal paralysis and of spasmodic tabes dorsalis and attributed it to a lateral sclerosis by reasoning from analogy only.

The first autopsies weakened this view.[24] Leyden from the beginning, Raymond from 1885 and till today 1898, deny this anatomo-clinical syndrome. This is also P. Marie's opinion (1892), who uses the word only for the infantile forms (Little's disease). I believe on the contrary with Brissaud (1895) that spasmodic tabes exists in the adult with a lateral sclerosis as the anatomical substratum. To constitute an anatomo-clinical syndrome such as this it is necessary that there be; clinically always the same symptom picture, anatomically a lesion of a constant location.

But aside from this constant lesion there may be other lesions, variable and clinically latent, without suppressing or altering the clearness of the type.

Thus Jean Charcot showed that Aran-Duchenne's progressive muscular atrophy exists without an amyotrophic lateral sclerosis even when the lesion is not strictly limited to the anterior horns of the grey substance. In order that the case become an amyotrophic lateral sclerosis it is necessary that the lateral lesion should be so important that it does not remain latent.

In the same way we may include in spasmodic tabes some cases of lateral sclerosis in which the le-

---

[24]PITRES a publié la première autopsie d'un tabes spasmodique de Charcot: c'était une sclérose en plaques.

sion extends slightly to some cellular groups, but without amyotrophy, or·to the tracts of Goll and to the cerebellar tracts without symptoms, or complicated by absolutely distinct lesions in the brain, for instance.

Applying these principles to the criticism of the facts published during the last years we will retain a great number of observations rejected by Raymond[25] and will say with Brissaud "By one of those revivals always necesssary it is now proved to us that a primary sclerosis of the lateral columns is not a myth. It really exists, and it is necessary to go back to it to find the cause and to conceive the pathogeny of a great number of cases of spasmodic tabes dorsalis." And as I have already remarked elsewhere, the same year (1898) where, in his third volume of clinics, Raymond said that the lateral theory of spasmodic tabes "has been completely destroyed," his interne Lorrain, under him, supported a thesis on family spasmodic paraplegia, which proves that the destruction of this theory is not complete. Then there is a second group of very clear cases in which there are clinically a pareto-spasmodic state and contractures, and anatomically lateral sclerosis.

3. Without retaining the more or less limited etymological sense of the word I reserve by the example of Brissaud and of van Gehuchten the name

[25]Tels sont notamment les faits de Strümpell (1879), Stoffela (1878), Morgan (1881), Aufrecht (1880), Minkowski (1884), Jubineau (1883), Westphal (1884).—Voir aussi la thèse de notre interne d'alors, Guibert (Montpellier 1892), les travaux de Jegorow (1891) et de Shüle (1894), les mémoires de Strümpell échelonnés de 1880 à 1894, les faits cités par Brissaud (1895), celui de Déjerine et Sottas (1896) et la Th. de Lorrain (Contrib. à l'étude de la parapl. spasm. familiale 1898).

of Little's disease for cases of spasmodic rigidity, observed in children born prematurely, without initial cerebral phenomena, and anatomically due to the absence of the development (at the time of birth) of the spinal portion of the pyramidal tracts. I eliminate thus from the group not only infantile spasmodic hemiplegias but all infantile cerebral diplegias.[26]

In these cases, thus defined, we find the clinical syndrome described above, and anatomically, not a lesion, but an absence of the pyramidal tracts.

C. Besides these three diseases the whole history of which is made up of the anatomo-clinical syndrome studied, we find this same syndrome (with others) in a certain number of other diseases.

Such is first disseminated sclerosis: In this disease the change in the antero-lateral columns is shown, 1st, by the characteristic tremor, an intention tremor, originating in action and exaggerated by the repetition of an act; 2nd, by the pareto-spasmodic phenomena which after the tremor constitute the most frequent and most characteristic symptoms of disseminated sclerosis.

When a diffuse myelitis or a compression of the cord changes, either directly or indirectly, by secondary descending degeneration the pyramidal fibres, we find also the syndrome we are studying. We discuss further along (in chapter 2) the cases of flaccid paraplegia in compression of the cord or in transverse myelitis.

---

[26]C'est le seul moyen de ne pas faire un groupe flou sans caractéristique clinique ou anatomique, et de répondre aux objections de Raymond, qui arrive à cette conclusion décourageante: "Les faits démontrent qu'à l'heure actuelle il nous est impossible d'établir un rapport fixe entre le mode de groupement et de localisation de ces symptômes et les lésions constituées à l'autopsie."

Certain symptoms of general paralysis (exaggeration of the tendon reflexes, tremor) are also dependent upon a change in this medullary system.

D. After all has been said and after what I have developed elsewhere, I believe it necesssary to keep the law I promulgated in 1877 and 1878 after Charcot and Strauss (1875) and to say, despite the opposition of certain authors, notably Raymond, that the permanent contractures of the pareto-spasmodic condition of medullary origin are in constant relation to a lesion of the spinal part of the pyramidal fibres.

But the pathological physiology of the syndrome remains obscure.

1. With Charcot, Vulpian and Brissaud[27] (1875 1880) it is necessary to think the permanent contracture due to a permanent muscular hyperactivity by an exaggeration of tonus. But, to explain the exaggeration of tonus, these authors admit that there must be a lesion of the pyramidal fibres, acting as strychnia, by exciting the root cells (center of tonus).

To the second part of their theory we may object: (A). Sclerosis of the pyramidal fibres should not have a special action on the cells; posterior sclerosis should have the same result; this clinically is not so; (B). We cannot comprehend the permanence of an exciting action exercised by a sclerosis without inflammatory activity; (C). In Little's disease one cannot comprehend that the absence of the pyramidal fibres excites the cells as a sclerosis of the same bundle.

It remains, then, to find how a lesion of, or the

----

[27]Je ne dis rien des théories de Folin et Hitzig qui sont réfutées partout.

absence of the pyramidal fibres produces this exaggeration of tonus shown by contracture.

2. Since Adamkiewicz (1881) it has been admitted that tonus is submitted to a higher regulating action formed by two antagonistic actions: The one inhibitory, which passes by the lateral columns, the other exciting, which the same author makes pass by the posterior columns.

From this notion the theory of Anton (1890) and of Pierre Marie (1892) follows.

The anterior root cell, the center of reflex tonus is a machine under pressure; an inhibitory action exercised by the higher centers normally arrives by the pyramidal fibres; when the pyramidal bundles are altered, destroyed or absent the inhibition is suppressed, the center free to excitation overacts; hence we have hypertonus and permanent contracture.

Here we have an advance in this theory over the preceding. But it is open to a serious objection (van Gehuchten): This theory does not explain that the symptomatology may be different when the lesion affects the cerebral portion or the spinal portion of this same pyramidal bundle, that in the first case there is paralysis, in the second contracture, that after a cerebral lesion the contracture appears only when the lesion descending has become subpontal and spinal.

This prime objection can be made by all the authors (Jackson, Bastian, Freund, Raymond) who locate in the brain (cortex) the origin of the inhibitory action transmittted by the pyramidal tracts to the anterior root cells.

3. Van Gehuchten[28] (1896, 1898) mentioned a

[28]J'ai essayé de montrer ailleurs que la théorie de Mya et Levi (1896), adoptée par Gerest (1898), ne résout pas non plus la question.

new element useful in the elucidation of this question. The inhibitory action of tonus comes from the higher centers, passes indeed by the pyramidal fibres but the exciting action passes by the indirect ponto-cerebello-spinal paths. Whence the pyramidal bundle is differently constituted in its cerebral and spinal portions: The lesion of the cerebral portion causes total paralysis; that of the spinal portion, involving only the inhibitory paths of tonus, brings about contracture. This explains very well the flaccid paralysis at the beginning of a cerebral lesion and the contracture of Little's disease or of lateral sclerosis at the onset, and here is the advance over preceding theories—but this does not explain the late contracture of the hemiplegic, it does not explain how the cerebral paralysis, flaccid at the beginning, becomes spastic when the lesion extends below and becomes sub-pontal.

Van Gehuchten understands the objection and responding to it admits that the late contracture of the hemiplegic is entirely different from the medullary contracture at the onset and from spasmodic contracture; in the hemiplegic the exaggeration of the tendon reflexes is related to the pathogeny of spasmodic contracture, but the permanent contracture which is thus separated from the exaggeration of the tendon comes simply from the fact that the extensors are generally more paralyzed than the flexors and so these less opposed by their antagonists contract.

Gerest very justly discussed this special theory of the contracture of hemiplegics: (A). In the extended cerebral softening there is not this unequal distribution of paralyses and yet contracture develops; (B). It is not understood why the contracture develops only late in hemiplegics; (C). In certain cases (neuritis for example) the paralysis can be very unequally distributed and yet contracture of the least paralyzed does not follow.

I add that it seems to me absolutely anti-clinical to separate the contracture of hemiplegics from exaggeration of tendon reflexes and from spasmodic contractures.

The pareto-spasmodic syndrome is always the same in its symptomatic expression and always corresponds to the same location of the lesion whether it is cerebral or spinal; a single thing distinguishes one from the other, that is the date of the appearance of the contracture; simply because contracture belongs solely to spinal and because the cerebral becomes spinal late, while the spinal is spinal from the onset. To dissociate the contractures of hemiplegics and the exaggeration of their tendon reflexes seems to me equally artificial and refuted by the clinic.

Among all the arguments given by van Gehuchten to oppose the contracture of hemiplegia to spasmodic contracture a single one is impressive: in the spasmodic form tonus is diminished. I might be contented to reply that this seems paradoxical with all the theories of contractures and in consequence with Babinski we must simply describe it as "singular." But we can reply more peremptorily. In a recent work on the question Marinesco (1898) concludes: "Even admitting that Babinski's conclusions have a general value this relaxation ordinarily exists in the non-paralyzed muscles, not in the contracted muscles.

"The result is that in no way should we conclude from Babinski's studies, as van Gehuchten has done, that the contracted muscles of the hemiplegic are found relaxed."

Then the second part of van Gehuchten's theory is not acceptable.

But, with only the first part of his ideas, we cannot respond to the objection formulated against all the theories which place the regulating center of tonus in the cerebral cortex. Then, we have not yet, despite all our accumulated efforts, a satisfactory theory of the relation between contracture and the pyramidal tracts.

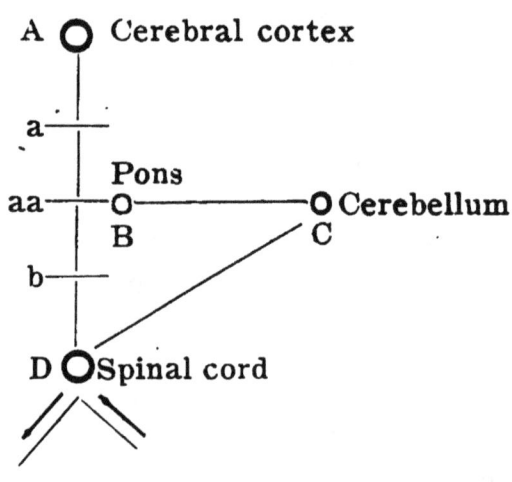

Fig. 1.

4. The failure of all these theories to refute the last objections is due to the fact that all place the origin of the inhibitory or controlling action exercised on tonus in the cerebral cortex. To remove all difficulties it is necessary and sufficient to place the higher center of the regulation of tonus not in the cerebral cortex, but lower in the pons.

Let us admit for a moment this hypothesis. Below in A (Fig. 1) is the cortical center of voluntary movements, which acts on the tonus when we wish to modify this reflex; in B (in the pons) is the center which rules automatic tonus, From this center

B (as from center A) direct fibres (by the pyramidal tract) go toward D (spinal center of reflex tonus) which carry inhibitory impulses, and indirect fibres (by the cerebellum C) which carry exciting actions.

When the lesion is located at a (cerebral portion of the pyramidal tract) there is motor paralysis: the orders given by A cannot reach D either by the direct or indirect fibres. But the tonus is not affected since its automatic center B remains in normal connection with D by the two classes of fibres, inhibitory and exciting. Then, no contractures.

When the lesion is located at b, that is to say the spinal portion of the pyramidal tract is attacked at the onset or finally, the tonus is no longer intact since B the automatic center of tonus no longer communicates with D by the inhibitory paths B D and still communicates with it by the exciting paths B C D.

It is indeed understood that the symptomatology differs according as to whether the lesion initially takes place above or below aa and that it also changes when the lesion initially above a finally invades the region below aa.

Here then is an hypothesis which answers all the objections: it consists simply in placing in the pons (B) and not in the cortex (A) the automatic regulating center of tonus.

This hypothesis is not unreasonable physiologically.

The cerebral cortex has certainly an action on the reflexes and on the tonus: the proof lies in the fact that we can modify voluntarily the attitude of our body, can act voluntarily on what Barthez calls the force of the fixed position, can to a certain limit control the sphincters.

But there is another thing: the complex reflexes as tonus have an automatic regulating center and it is from this center that the inhibitory and exciting actions we are studying arise.

This automatic center is entirely distinct from the voluntary center (cortical) as it is distinct from the lower simple reflex center (spinal)..

We maintain some attitudes, even complex ones, entirely without voluntary action and higher consciousness.

It is this automatic center that I locate in the pons. The physiologists have made some experiments which seem to establish this point.

Vulpian says, "It is the pons which presides over the normal attitude of animals." He shows a very young rabbit and a pigeon from which all portions of the encephalon anterior to the pons have been removed, and which hold themselves in the normal attitude, raising themselves if they are put on the back or on the side. A fowl thus operated upon can hold itself on one foot or hide its head under its wing. More recently Goltz has shown a frog without out a brain doing "acrobatic exercises; if it is put on a plank which gradually inclines it climbs and passes over from one side to the other without falling." Two dogs, from which Goltz had extirpated the greater part of the cerebral hemispheres "were essentially reflex machines eating and drinking." Hedon who reported some experiments, concludes,[29] "that animals deprived of the brain preserve besides organic functions which remain intact, various faculties which may be classified under the titles of equilibration, of coordination of movements and of emotional expression."

These various experiments show that the regulating center of the attitude is neither in the cord nor

[29]HEDON. Précis. de physiol. 2e édit. 1899, p. 511.

the brain. For if, in the frog above cited, the mutilation is extended and a part of the bulb is removed the tendency toward the normal attitude immediately is rendered impossible. Hence the center of attitude is located in the pons or at least in the mesencephalon.

But some say (Brissaud) that the contractures as in strychnia poisoning, only exaggerate and immobilize the attitude. We can say then that from this pontal (or mesencephalic) center a double regulating action on tonus arises, an inhibitory by the pyramidal fibres, and an exciting by the ponto-cerebellar fibres.

Then the contracture of spinal origin is allied to an alteration or absence of the pyramidal fibres, this alteration determining the contracture by the suppression of the inhibitory action on tonus which arises from the pons and goes to the root-cells by the pyramidal fibres.

E. The differential diagnosis needs to be made only from neuroses and cerebral lesions.

For the neuroses, paralysis agitans with a tremor (in repose) has entirely different characters from that of disseminated sclerosis; hysteria has its stigmata and its other peculiar symptoms and often has contractures, but mobile, fleeting ones, usually without exaggeration of the tendon reflexes and without clonic phenomena.

When the contractures are of cerebral origin, they are early if from a local lesion, or accompanied by other symptoms either of a tumor or of an acute or subacute encephalitis.

Tetanus or strychnia poisoning and intoxications with contractures all have a special clinical history which easily distinguishes them from affections of the medullary lateral system we are studying.

3. The Associated Syndrome of the Posterior and Lateral Columns: Ataxo-spasmodic State.

I called attention (after others) to this anatomo-clinical syndrome in 1886[30] and the very title of my work shows I did not wish to confound this that I called "combined tabes" with what has been called "combined or associated sclerosis" of the cord. This latter term, purely an anatomical one, is applied to a great number of separate cases forming hence a confused group.

In proportion as the histological technique has been perfected more and more accessory lesions have been discovered by the side of the principal lesion and then studying the combined sclerosis we conclude, as P. Marie, that they are a diffuse group having no separate existence.

I believe that it is good, on the contrary, to reserve the name of combined tabes for the single complex anatomo-clinical syndrome formed by the superposition in the same subject, not of two lesions, but of two well defined anatomo-clinical syndromes: ataxic tabes with its posterior sclerosis, and spasmodic tabes with its lateral sclerosis; that is to say ataxo-spasmodic tabes with postero-lateral sclerosis. Thus defined, the group is very clear, well characterized, as all the anatomo-clinical syndromes, at the time by a fixed location of the lesion.

This is admitted by Brissaud who finds "detest-

---

[30]Tabes combiné (ataxo-spasmodique) ou sclérose postérolatér. de la moelle. *Arch .de neurol.* 1886, t. XI, p. 156 et 380; t. XII, p. 27.—Voir aussi TARBOURIECH. Contrib. à l'étude du tabes combiné. Th. de Montpellier 1888, no 83, et GUIBERT. Et. clin. de la sclér. primit. des cord. latér de la moelle. Th. de Montpellier, 1892, no 23.

able" the word combined tabes but admits what he calls "ataxo-spasmodic paraplegia" if not as a special at least as a "definite nosographical variety."

We do not ask more.

Combined tabes is the disease which is exclusively manifested by this anatomo-clinical syndrome.

Besides this other diseases may also have this syndrome, but in the midst of others. I would cite without insisting upon them, disseminated sclerosis, general paralysis, diffuse myelitis and also medullary arteriosclerosis on which P. Marie has insisted with reason.

There is nothing to say of the pathological physiology contained in the two preceding paragraphs nor of the differential diagnosis of which we have also analyzed the constituent elements.

4. The Syndrome of the Anterior Horns: Muscular Atrophy.[31]

A. In two important groups of cases the lesions are systematized exclusively to the anterior horn cells: progressive muscular atrophy (type of Aran-Duchenne) acute spinal, atrophic paralysis (infantile and adult). We add to these certain cases of experimental myelitis.

1. Clinically discovered by Duchenne (1849) and by Aran (1851) progressive muscular atrophy, had its pathological anatomy fixed by all the French school from Cruveilhier (1853) to Haven, Charcot and Vulpian.

Then from this classification, initially too large, there have been successively removed the progres-

---

[31]Voir nos leç. sur trois cas d'atr. muscul. L'atr. muscul. est le syndr. du neurone moteur central (bulbomédull.) inférieur, in Leç. de clin. méd. 1898, 3e série. p. 793.

sive myopathies, neuritis, amyotrophic lateral sclerosis, syringo-myelia and P. Marie has finished by writing, "The progressive muscular atrophy of Duchenne of Boulogne[32] does not belong there."

This was an exaggeration.

As we have said above à propos of lateral sclerosis you can retain in a given classification some cases whose lesion is not strictly limited to a system provided that the lesion outside of the system is accessory, of little importance, and above all clinically silent.

On this principle, we find the thesis of Jean Charcot,[33] observations of Strumpell, Oppenheim, Dejerine, Darkchewitsch and a personal experience (plus one more recent of Raymond)[34] which establish histologically the existence of this syndrome and its lesion. Hoffman[35] equally has shown that in the hereditary amyotrophies a certain part should be classed as the Aran-Duchenne type.

Here then is the first group of cases the existence of which remains indisputable, characterized, clinically, by the muscular atrophy, anatomically, by a lesion of the anterior horns of the grey substance.

2. It was still Duchenne who fixed the clinical

[32]PIERRE MARIE. Existe-t-il une atr. musc. progr. Aran-Duchenne? *Revue neurol.* 1897, t. V, p. 686.

[33]JEAN CHARCOT. Contrib. à l'ét. de l'atr. musc. progr., type Aran-Duchenne. Th. Paris 1895.—Voir aussi TZEDEPOGOLU. Th. Montpellier 1892.

[34]RAYMOND. Clin. des mal. du syst. nerv. 1897, 2e série, p. 449.—Voir aussi TARGOWLA. Un Job moderne, atr. musc. du type Aran-Duchenne chez un chemineau. *Nouv. Inconog. de la Salpêtr.* 1897, t. X, p. 415.

[35]HOFFMAN. Ueb. d. progress. spin. Muskelatr. in Kindesalter aus. famil. Basis. Deutsches Zeitschr. f. Nervenh. 1893, t. III, p. 427 et Weit. Beitr. z. Lehre von d. heredit. progr. spin. Muskelatr. im Kindesalter. *Ibid.* 1897, t. X, p. 292. (Trav. de la clin. d'Erb et du labor. d'Arnold).

description of *infantile spinal atrophic paralysis*,[36] already seen by Underwood (1774), Heine (1840), Rilliet and Barthez (1851): after an acute period of generalized paralysis the disease is localized in some muscular groups and is manifested exclusively by an atrophy.

Vulpian and Prevost (1866) were the first to localize the lesion in the anterior horns of the grey substance. A localization confirmed by Lockhart, Clarke (1867), Charcot and Joffroy (1870).

Duchenne had already seen that the same disease may develop in the adolescent and the adult, and all the later works have confirmed the existence of an acute spinal paralysis of the adult, clinically and anatomically identical with that of the infant.

3. Experimentally[37] with old cultures of the streptococci of erysipelas Rogers (1891) was "able to reproduce in six animals a system myelitis characterized from an anatomical point of view by a degeneration of the anterior horn cells; from the symptomatic point of view by an ensemble of phenomena comparable to progressive muscular atrophy."

Some analogous results have been obtained by Gilbert and Lion (1891) with the coli-bacillus, Bourges (1893) with the erysipelococcus, Thoinot and Masselin (1894) with the coli-bacillus and the staphylococcus.

B. But it does not seem necessary to dwell on the diseases which attack the anterior horns without being exclusively localized there: it suffices to say

---

[36]Voir les thèses de DUCHENNE fils, Montpellier, 1864, nº 8; et LABORDE, Paris 1864, nº 163.

[37]Voir notre Rapport au Congrès de Bordeaux sur les Myel. infect., in Leç. de clin. méd. 1898, 3e série, p. 540.

3

that when the anterior horns are thus attacked, clinically there will be muscular atrophy.

This is what happens notably in certain complicated cases of tabes, in certain syringo-myelias, in certain old hemiplegias with descending degenerations, in certain abnormal cases of disseminated sclerosis, in pachymeningitis.

C. The synthesis of the syndrome of the anterior horns is all made since it is reduced to amyotrophy, and the pathological physiology does not require discussion: it is the expression of a trophic action exercised on the muscle by the cell bodies of the central inferior motor neurone.

D. Differential diagnosis. It is necessary first to eliminate the neurosic origin of amyotrophy when it is met.

Since Babinski's memoir[38] the facts concerning hysterical muscular atrophy have been multiplied but in these cases (very rare, moreover), there are no fibrillary twitchings, no reaction of degeneration, and there are other symptoms of hysteria.

The truly diagnostic difficulty consists in distinguishing spinal amyotrophies from neuritic amyotrophies, and from myopathic amyotrophies. (The brain by itself does not cause muscular atrophy).

In neuritic amyotrophies the distribution of the atrophy is that of one or more nerves, the paresis or paralysis is more marked, there are pains (spontaneous and on pressure of the nerve trunks), the tendon reflexes are diminished or lost, the etiology is a special one and retrocession frequent.

In myopathies there are no fibrillary twitchings, tendinous retractions are frequent, no reaction of

[38]BABINSKI. De l'atr. muscul. dans les paral. hystér. *Arch. de neurol.* 1886. Voir GILLES DE LA TOURETTE. Traité clin. et thérap. de l'hyst. 1895, t. II, p. 503.

degeneration, begins more frequently at the extremity of a limb, often there is a family history of the same.

5. The Associated Syndrome of the Lateral Column and the Anterior Horns: Spastic Muscular Atrophy.

From 1865 Charcot observed a sclerosis of the lateral tracts in amyotrophy and with Joffroy (1869-1879), described amyotrophic lateral sclerosis (Charcot's disease) which is thus clinically defined: "a progressive paresis of certain muscles, at times followed by atrophy and more often by contractures of these muscles or by phenomena analogous to this contracture" as the rigidity, more or less difficult to overcome, the tremor in certain cases; then we should add the exaggeration of the tendon reflexes and the clonic phenomena.

Anatomically there is in the cord (let us not concern ourselves here with the medulla nor with the brain) a lesion of the anterior horns as in progressive muscular atrophy and an alteration of the pyramidal fibres as in descending degenerations. P. Marie[39] and Brissaud have shown that there is also a lesion of the mass of the antero-lateral fibres and especially of the short fibres called the fibres of the cells of the cord, which put the various segments of the cord in relation with each other. This lesion of the intramedullary relay neurones plays perhaps a considerable role, if not an exclusive one (Brissaud, Gerest) in the production of amyotrophic lateral sclerosis. This completes, without disturbing, the

---

· [39] PIERRE MARIE. Leç. citées, 1892 et *Soc. méd. des hôp.* 1893, déc.

first notion of the anatomo-clinical syndrome we are studying, the pyramidal lesion remaining moreover "much the more salient" (Marie).

A single case constitutes a real exception in this group; it is that of Senator (1894): a clinical picture of amyotrophic lateral sclerosis and no lateral sclerosis on autopsy. But this observation remains isolated and is incomplete since the brain was not examined.

If amyotrophic lateral sclerosis manifests itself exclusively by this anatomo-clinical syndrome this same syndrome may also be found, more or less as an episode in other maladies as diffuse myelitis, disseminated sclerosis,[40] etc.

There is nothing to say of the pathological physiology and the differential diagnosis which has not been said in paragraphs 2 and 4.

6. The Syndrome of the Centro-posterior Grey Substance: Dissociation of Sensation called Syringo-myelic (and Vaso-motor Troubles).

A and B. This syndrome has been studied in syringo-myelia, where perhaps it is found most frequently.

But it is certainly also found in other cases and the expression syringo-myelic, for this dissociation of sensibilities preserves an error if taken literally.

Named by Ollivier of Angers (1827) syringo-myelia is an anatomical condition (known since 1688) characterized by the presence in the cord of abnormal cavities. We say there is a hydro-myelia

---

[40]Voir notamment le fait récent de BRAUER. *Neurol. Centralbl.* 1898, p. 638. et *Revue neurol.* 1899, p. 22.

(the analogue of hydrocephalus) when there is a dilation of the central canal.

Developing Grimm's ideas (1869) Simon (1874), and Leyden (1876) Roth[41] in a series ot works since 1882 and Dejerine[42] in 1889 have attributed syringo-myelia exclusively to a medullary gliosis and have considered the two expressions as synonyms. But giving the ideas of Hallopeau (1870) in detail and of Charcot and Joffroy (1869) Joffroy and Achard[43] and others have shown that syringo-myelic cavities can have a myelitic origin also, and Souza Martins[44] has shown that it may equally be observed in lepra. On the other hand, medullary gliomatosis may evolve without producing cavities. The synonymousness then of the two must be refused and the word syringo-myelia must keep its original anatomical meaning.[45]

The principal symptom of syringo-myelia is the following, dissociation of sensation, analgesia and thermo-anæsthesia with preservation of tactile sensation. Kahler[46] and Schultze[47] first diagnosticated a syringo-myelia by this symptom. The thing became classic.

With Roth and Dejerine this syndrome has been

[41]ROTH. *Arch. de neurol.* 1887-89, nos 42 et suiv.

[42]DÉJERINE. *Soc. méd. des hôp.*, 22 févr. 1889.

[43]JOFFROY et ACHARD. De la myel. cavitaire. *Arch. de physiol.* 1887.

[44]SOUZA MARTINS. Un caso de syringom. dépend. della labra. Anal. in *Revue neurol.* 1894, p. 307.

[45]BRISSAUD, qui a bien étudié la dissociation "syringomyélique" dans la pachyméningite cervicale hypertrophique (Leç. sur les mal. nerv. 1895, p. 196) et cite les cas avec autopsie de Joffroy, de Pierret et de Kœler, les considère comme des exemples de gliomatose consécutive, secondaire.

[46]KAHLER. *Prag. med. Wochenschrift*, 1882 et 1888.

[47]SCHULTZE. *Virch. arch.* 1882, et *Zeitschr. f. klin. Med.*, t. XIII.

completely limited to this lesion and the qualification "syringo-myelic has been given to this dissociation of sensation.

Thus this symptom has become an absolute sign of syringo-myelia superior even to the anatomical facts. Thus Dejerine insisted to Joffroy that he had observed a true syringo-myelia (although it was at the autopsy) unique because the patient had not presented the dissociation called syringo-myelic.

I believe that I was one of the first[48] to protest (1889) against this view which reversed all that is known of the semeiology of the nervous system. All the symptoms known are in relation with the location of the change; only this would have expressed not the location but the anatomical nature of the change. This would be improbable.

To support my protest I have collected two orders of proofs which I will recall, adding to them the documents prepared since then, which have singularly confirmed and rendered definite the thesis under discussion.

1. *Syringo-myelia can exist without the syndrome of dissociation.*

I have cited a personal experience observed with my colleague Carrieu. Since, out of 66 cases with clinical and autopsy observations, referred to in Anna and Baumler's thesis (1887) I found 55 in which there was neither the entire syndrome nor any of the constituent elements of the syndrome (and this is also in gliomatosis and lacunar myelitis).

Since then proofs of this first proposition have

[48]Voir nos leç. sur le Syndr. bulbomédull. constitué par la thermanesth., l'analg. et les troub. sudor. ou vasomot. (subst. grise latéropostér.), in *Montpellier méd.* août 1889 et Leç. de clin. méd. 1891, 1re série, p. 186.

accumulated. First there is the observation of Joffroy and Achard[49] on syringo-myelia demonstrated on autopsy and having produced total anaesthesia without any dissociation.

Then Rosenblath[50] published two cases of syringo-myelia (with autopsy) without sensory troubles or at least without dissociation; Préobrajensky,[51] a case of syringo-myelia, not gliomatous: total anaesthesia without dissociation. Dimitroff[52] began an important work on syringo-myelia with an observation of his own without sensory troubles. Déjerine and Thomas[53] have observed a recent case without sensory troubles.

In the great monograph of Schlesinger[54] (1895) may be found a number of details on abnormal types of syringo-myelia and finally Raymond[55] has many times insisted on the clinical polymorphism of this

[49]JOFFROY et ACHARD. Un cas de mal de Morvan avec aut. *Arch. de méd. expérim.* 1890, p. 540.

[50]ROSENBLATH. Z. Casuist. d. Syringom. etc. *D. Arch. f. klin. Med.* 1893, t. LI, p. 210 (*Revue neurol.* 1894, p. 11). Trav. de la clin. méd. de Leipsig.

[51]PREOBRAJENSKI. Mem. méd. 1894. (*Revue neurol.* 1895, p. 75).

[52]STEPHAN DIMITROFF. Ueb Syringom. (Trav. de la clin. méd. du prof. Eichhorst à Zurich), *Arch. f. Psych.* 1896, t. XXVIII, p. 582 et 1897, t. XXIX, p. 299: fait suite au travail d'Anna Baümler, et analyse 297 observ. (jusqu'en octobre 1891).

[53]DÉJERINE et THOMAS. Un cas de syringom. type scapulohumér. avec intégr. de la sensibil., suivi d'aut. (intégr. de la subst. grise médiane). *Soc. de biol.,* 10 juillet 1867, (*Revue neurol.* 1898. p. 153.

[54]SCHLESINGER. Die Syringomyelie (trav. de la 3e clin. méd. et de l'Institut d'anat. et de physiol. des centr. nerv. à l'Univers. de Vienne). Leipsig et Vienne 1895. (526 indicat. bibliogr. classées par ordre alphabét d'auteur).

[55]RAYMOND. Clin. des mal. du syst. nerv. 1896, t. I, p. 315 et 327, et 1897, t. II, p. 580 et 516.

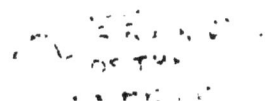

disease. This last author insists especially on the frequency of the total anaesthesia (non-dissociated) in syringo-myelia and cites the findings of Miura (1889), Rumpf (1889), Hochhaus (1891), Homen (1894), Oppenheim (1893), Schuppel. He cites also some reports of syringo-myelia in the form of disseminated sclerosis (Brutton, Rosenblath) then reports a form of spasmodic tabes: Strumpell (1880), Kahler (1882), Reisinger and Marchand (1884), Schlesinger (obs-7), Raymond (1893).

The demonstration of our first proposition is then well established: syringo-myelia can exist without the syndrome of dissociation.

2. The dissociation syndrome can exist without syringo-myelia.

Morvan described (1883), under the name of analgesic panaritia, paresis of the upper extremities, a disease which bears his name and in which one often finds the dissociation syndrome. Gombault and Reboul (1889) made an autopsy on one of the patients of Morvan with dissociation. They found a neuritis, a myelitis, but no medullary lacuna.

In 1889 I declared the principle, although unique, and expressed the conviction that "Similar cases existed and would multiply themselves when attention was directed to this point." This has been largely realized.

First Charcot[56] found syringo-myelic dissociation in hysteria, in lepra (Leloir, Babinski, Thibierge) and in compression of the cord[57] or of nerves (Critzman). Minor[58] described it in many

[56]CHARCOT. Leç. du mardi 1889, p. 517.
[57]CHARCOT. Un cas de pseudosyringom. in *Sem. méd.* 1891, p. 193.
[58]MINOR. De l'hématom. centr. *Soc. de neurol. et de psych. de Moscou,* 17 déc. 1893 et Rech. clin. et anat. sur

cases of traumatic haemato-myelia.[59]   Freund[60] carefully studying sensory troubles in disseminated sclerosis, described "syringo-myelic" dissociation in ten observations of this disease (with peculiar character of less duration than in true syringo-myelia).

In 1892 àpropos of a personal case at Brissaud, Jean Charcot[61] studying "the dissociations called syringo-myelic in compressions and section of nerve trunks," recalls the fact that diverse dissociations are found in diseases of the skin,[62] (psoriasis, eczema) and again finds true dissociation in a series of cases of section or compression of nerves (Letievant, Weir Mitchell, Richet, Chaput, Blum).

More recently Cavazzani and Manca[63] have found this same dissociation after a traumatic section of the radial nerve. And the same year (1895) Brissaud[64] was able to say: "The clinic has instructed us on the pretended pathognomonic value of syringo-myelic dissociation."

This dissociation of sensibility no longer pertains to a single disease." And he cites a case of Brown-Sequard with dissociation cured by anti-

les aff. traumat. de la moelle suivies d'hématom. centr. et de format. cavitaires centr. Congrès de Moscou 1897 (*Revue neurol.* 1894, p. 212 et 1897, p. 549).

[59]Voir VERCHERE. De la parésie analgés. à panaris. des extrém. super. De la parésoanalg. des extrém supér. ou mal de Morvan. *Revue des sc. méd.* 1891, t. XXXVIII, p. 324.

[60]FREUND (de Breslau). Ueb. d. Vork. von Sensibilitätsstör bei. mult. Herdskler (trav. de la clin. du prof. Westphal). *Arch. f. Psych.* 1891, t. XXII, p. 317 et 588.

[61]JEAN CHARCOT. *C. R. de la Soc. de Biol.* 1892. p. 941.

[62]RENDU. Les anesth. spont. Th. d'agrég. Paris, 1875.

[63]CAVAZZANI et MANCA. Alteraz. della sensibil. termice e tattile in seg. a les. del nervo rad. *Rif. med.* 1895, n° 57 (*Revue neurol.* 1895, p. 534).

[64]BRISSAUD. Leç. sur les mal. nerv. 1895.

syphilitic treatment and in consequence without syringo-myelia.

In 1896 Max Laehr[65] reported a series of facts of "syringo-myelic" dissociation in some cases of the Brown-Sequard syndrome without syringo-myelia, especially those of Muller (1871), of Charcot and of Gombault (1873) Gowers (traumatic hemorrhage) Beevor (syphilitic tumor), Steel and Williamson (1893).

The same year Schlesinger[66] published a case (and calls attention to three other personal cases) of intramedullary tumor (without syringo-myelia) with dissociation.

Hanot and Henri Meunier[67] observed an analogous syphilitic gumma. There is also a case of dissociation by syphilitic meningo-myelitis which Piaott and Cestan[68] published, while David Edsall[69] studied the dissociation of sensibility of a syringomyelic type in Pott's disease, and which James Hen-

---

[65]MAX LAEHR. Ueb Störr. d. Schmerz-Temperatur. Empfind. in F. von Erkank. d. Rückenm. (Klin. St. mit besond. Berücksicht. d. Syringom.) Trav. de la clin. du prof. Jolly. *Arch. f. Psych.* 1896, t. XXVIII, p. 773.

[66]SCHLESINGER. Tumeur médull. (Gliosarc. de la rég. cervic.) Club méd. Vien., 22 janvier 1896. (*Revue neurol.* 1896, p. 441).

[67]HANOT et HENRI MEUNIER. Gomme syphilit. double de la moelle épiniére ayant déterm. un syndr. de Brown-Seq. bilat. avec dissociat. syringomyél. *Nouv. Inconogr. de la Salpêt.* 1896, p. 49 (même cas commun. par Henri Meunier au Congrès de l'A. F. A. S. à Carthage 1896.)

[68]PIATOT et CESTAN. Syndr. de Brown-Seq. avec dissociat. syringom. d'orig. syphilit. *Ann. de dermatol.* 1897, p. 713 et *Revue neurol.* 1897, p. 645.

[69]DAVID EDSALL. Soc. neurol. de Philadelphie, 20 déc. 1897. *The Journ. of nerv. a. ment. Dis.* 1898, p. 257 et *Revue neurol.* 1898, p. 742.

dried Lloyd[70] demonstrated in traumatism of the cor .

In a work inspired by Marinesco, Vines[71] shows "that syringo-myelic dissociation is not a rare symptom in various cases of chronic myelitis." Dejerine and Thomas[72] have observed dissociation in a case of hypertrophic gummatous pachymeningitis.

And Raymond[73] recently has written: "It is not very long ago that dissociated anaesthesia was thought to pertain properly to syringo-myelia. A sort of pathognomonic signification was attributed to it. Today it is recognized that this variety of anaesthesia is observed in very diverse pathological circumstances. It seems natural this should be so.

If a gliomatous tumor or any other neoplasm can produce these symptoms by interrupting the conductors of sensibility it is easily conceived that the lesion may be of any nature, a hemorrhage, focus of softening, or an islet of sclerosis,[74] the

---

[70]JAMES HENDRIE LLOYD. Traumat. aff. of the cerv. reg. of the Spin. cord. simulat. Syringom. *The Journ. of nerv. a. ment. dis.* 1894. p. 343 (*Revue neurol.* 1894, p. 450) : histoire clinique de deux cas.—A Study of the los. in a case of trauma. *Brain.* 1898 (*Revue neurol.* 1898, p. 613) : autopsie de l'un des deux faits précédents.

[71]VINES. ·Despe dissociationen. *Romania medicala,* 1898 p. 122 Anal. par Marinesco in *Revue neurol.* 1898, p. 850.

[72]DÉJERINE et THOMAS. Un cas d'hémiparapl. avec anesth. croisée. Syndr. de Brown-Séq. suivi d'aut. *Arch. de physiol.* 1898, p. 594.

[73]RAYMOND. Clin. des mal. du ςyst. nerv. 1897, t. II, p. 549.

[74]C'est la pensée qu'ont également énoncée Hanot et Meunier, dans leur travail déjà cité de 1896 : "...Quoi de plus naturel, dès lors, qu'une pareille lésion, équivalant physiologiquement à une lacune syringomyélique ait déterminé la dissociation des sensibilités, telle qu'on le recontre dans les cas de gliomatose médullaire?"

symptoms will be the same in all cases as facts prove."

We can then affirm as demonstrated today the proposition which we supported in 1889 on a single disputable case: *the syndrome dissociation can exist without syringo-myelia.*

C. From all that precedes it results that the anatomical unity of the dissociation syndrome is not made by the presence of medullary gliomas, nor by the presence of medullary cavities nor by the presence of a single constant medullary lesion, but that for this syndrome as for all other medullary symptoms, the unity is only made by the unity of the location of the lesion. It is necessary then to seek now what the unique and constant location of the lesion is corresponding to this dissociation syndrome. We shall thus come to the pathological physiology of this syndrome.

In the medullary sensory system it is not a lesion of the posterior columns which gives rise to our syndrome. For in the syndrome of the posterior columns (which we have already studied above) there is total anaesthesia or if there is dissociation it is tactile anaesthesia and especially muscular which dominates with persistence and often exaggeration of the sensation of pain and temperature. This is then an inverse dissociation and in a manner complementary to the dissociation called syringo-myelic. Further, in many observations with very distinct "syringo-myelic" dissociation, the integrity of the posterior columns is noted and in observations of syringo-myelia with total anaesthesia there has been noted on the contrary a participation of the posterior columns in the lesion.

In every case of syringo-myelic dissociation the alteration is in the *posterior horns of the grey matter.*

We have indeed cited some cases in which the dissociation appeared to respond to a lesion of the peripheral nerves. But then it is permissible to suppose that the neuritic lesion has extended to the medullary centers as in Marinesco's[75] case. It may also be said that "There are distinct points on the skin for tactile and thermal sensibility."[76] But for our study it is sufficient to say that *when the so-called syringo-myelic dissociation is due to a medullary lesion, the change is in the posterior horns of the grey matter.*

This is the clinical conclusion which is the same as in 1889.[77] *Physiologically* the question has advanced less.

Moreover, I can cite only the celebrated experiment of Schiff which has not been repeated: he made a complete section, except the posterior columns of the dorsal cord of a rabbit. The animal preserved touch sensation very distinctly and completely lost sensations of heat and pain.

In 1894 Holzinger[78] undertook the following experiment in the clinical laboratory of Bechterew of St. Petersburg. He cut the dorsal cord of dogs (at

---

[75]MARINESCO. Les polynévr. en rapport avec la th. des neur. *Soc. de Biol.*, 30 nov. 1895. Voir aussi *Presse méd.* 28 déc. 1895 et Rapport au Congrès de Moscou sur l'histopathol. de la cell. nerv., août 1897 (*Revue neurol.* 1896, p. 54 et 70; 1897, p. 523) : "Par conséquent, conclue-t-il, il n'existe pas de névrites sans réaction des cellules des nerfs atteints."

[76]HEDON. Précis. de physiol. 1896. Voir dans ce livre (p. 474) les fig. de Goldscheider.

[77]"En un mot, on peut dire que l'altération des cornés postérieures a pour traduction symptomatique le syndrome sur lequel je me suis longuement étendu." Leç. de clin. méd., 1re série, p. 243.

[78]BECHTEREW. Die sens. Bahnen in Rückenm. *Neurol. Centralbl.* 1894, p. 647.

the level of the 3rd and 4th pairs of nerves) and noted: 1st. For sensation to pain (A) hypaes-thesia, transitory (some days) bilateral on section of the lateral half of the cord; (B) nothing, on si-multaneous section of the posterior columns, of the grey matter and of the anterior columns, and of the anterior part of the lateral columns with a part of the anterior horns; (C) analgesia of all the body below level of lesion by the section of the two lateral columns or by section of the posterior half of the cord; only in this latter case the section must be a little before (anterior to) the pyramidal tracts; 2nd. For tactile and muscular sensations, anaes-thesia when the posterior columns are destroyed (and then also ataxia). Some analogous results con-firmative of the same conclusions appear to have been obtained by Münzer and Wiener.[79]

From these physiological and especially these clinical reports the probable path in the cord of the fibres conducting pain and temperature[80] may be deduced.

All conducting fibres enter by the posterior root and penetrate the posterior horns of the same side at once: whence we have thermic and pain anaes-thesia in certain cases of a lesion limited to the pos-terior horns; analgesia of the *same side* as the le-sion.[81] In these cases the analgesia and thermo-

---

[79]Münzer et Wiener. Sur la destr. isolée de la subst. grise médull. *Arch. f. experim. Pathol. u. Pharmak.,* t. XXXV, p. 113. et *Revue neurol.* 1895, p. 586.

[80]Voir notamment sur ce point le travail déjà cité de Max Laehr (1896) et celui de Schlesinger. Localisat. d. Schmerz u. Temperatursinnsbahnen in Rückenm. Wien. physiol. Club 26 mars, *Neurol. Centralbl.* 1895, p. 751.

[81]Tel est le cas présenté (avec autopsie) par Déjerine et Sottas à la Société de Biologie le 23 juillet 1892 (p. 716 des C. R.).

anaesthesia are *segmental,* each segment of the limb having its special medullary center.[82]

Then these conducting fibres pass to the opposite side (by a commissure) : whence we have *crossed* analgesia when the lesion is above the level of the center of the region affected (the case in Brown-Sequard syndrome). This intra-crossing does not seem to take place at the same height for the different fibres, thermic and algesic of the same region; for, in a lesion limited to the posterior horns, the two peripheral zones of anaesthesia are not necessarily superimposed. After their intra-crossing these conducting fibres pass into the grey matter and very probably pass at once into the sensory bundles of the anterior lateral columns (notably into Gower's bundle).

As to tactile and muscular impressions, they are not obliged to pass by the relay neurone (posterior horn) of their region. When this neurone is destroyed they continue either by the relay neurones of a higher level or by the posterior white substance alone and thus reach the brain; whence we have the so-called syringo-myelic dissociation of sensation in a part of a limb of the same side as the lesion when the lesion is located in the corresponding region of the posterior horn.

Above this relay neurone thermic and pain impressions continue by the grey substance and the lateral columns, while tactile impressions follow another path (posterior columns). From this one can understand that a lesion limited to this level can produce so-called syringo-myelic dissociation of the opposite side as compared with the lesion if located

[82]Voir nos Leç. sur les Sympt. médull. à distrib. segment. publiées par mon chef de clin. le Dr Gibert dans le *Nouveau Montp. médical* (1899).

at one point, and inverse or complementary (tabetic) dissociation,[83] if seated at another point.

Then, to resume, from pathological physiology it is perceived that *the dissociation called syringomyelic is not exactly the syndrome of the posterior horns,*[84] *but often the syndrome of the second sensory neurone, (the first neurone of the relay), the cellular bodies of which are in the posterior horns, and whose cylinder axis prolongations are in the lateral column and especially Gower's bundle of the opposite side;* that is to say, that the dissociation is on the same side as the lesion when this attacks the cellular bodies of this neurone and that the dissociation is on the side opposite the lesion when this involves the prolongations of this same neurone.[85]

D. To make the differential diagnosis of this syndrome it is sufficient, as soon as the dissociation is established, to discuss its origin; hysterical, neuritic, or medullary.

In hysteria the distribution may be segmental; but there are other symptoms which permit a diagnosis of the neurosis. It becomes more difficult when there is a hystero-organic association. Sug-

---

[83]La dissociation inverse ou complémentaire est constituée par l'anasthésie tactile et musculaire avec conservation et souvent exaltation de la sensibilité à la température et à la douleur.

[84]Pour Marinesco (*Soc. méd. des hôp.* 6 *mars* 1896), la dissociation syringomyélique est due à l'interruption du contact utile entre les collatérales et certains neurones de la corne postérieure; ce phénomène d'addition et de renforcement n'ayant plus lieu quand la corne postérieure est détruite.

[85]Au moment de mettre sous presse, nous recevons un travail de van Gehucten sur "la dissociation syringomyélique de la sensibilité dans les compressions et les traumatismes de la moelle épinière, et son explication physiologique." (*Sem. méd.* 1899, n° 15, p. 113).

gestion here seems to be the only absolute means of resolving the question.

As to the peripheral origin I believe this rare without medullary involvement. On the other hand the dissociation is then always distributed according to the nerve territories and not in segments; and there are all the other ordinary symptoms of neuritis.

E. It is now time to consider that sudorific and vaso-motor disturbances make a part of the syndrome of the posterior horns.

I will devote to them only an appendix to the general chapter as their physiological and semeiological history are not complete.

In a patient whom I studied in 1889 there were more exaggerated sweats and hyper-thermia on the same side as the thermo-anaesthesia and the analgesia. I have collected 29 cases, borrowed from various authors,[86] in which there was with the dissociation of sensibility on one part, sudorific or vaso-motor troubles on the other.

From this frequent association of the two classes of phenomena we have already inferred that probably the lesions corresponding to the symptoms have the same location or both locations are near.

The physiologists do not appear to know anything of this location of the medullary vaso-motor center (or of the great sympathetic).

Some make the anterior columns intervene (Brown-Sequard, Schiff) or the column of Clarke (Jacubowitz, Gaskell), but without proving their convictions.[87]

[86]STRUMPELL, FÜRSTNER et ZACHER, GLASER, REMAK, SCHULTZE: WILL, GULL, LOCKART-CLARKE, et HUGHLINGS-JACKSON, WESTPHAL, KRAUSS, OPPENHEIM, FREUND, ROTH, MORVAN, PROUST, DÉJERINE, RUMPF, MOUTARD-MARTIN.

[87]Voir les art. Vasomoteurs de NUEL et Sympathique de FRANCOIS FRANCK in Dict. encyclop. des sc. méd.

4

The hesitation of the physiologists is still greater for the localization of the medullary center of the sudorific nerves.[88]

In the clinic, Bouverat showed hyperidrosis in various cord diseases, but without saying anything as to the location of the lesion. In 1882 Pierret[89] established much more, and by applying the anatomo-clinical method to the study of the vaso-motor troubles of tabes he places the lesion in the intermedio-lateral tract of Clarke.

This is where we were in my lessons in 1889.

A little later the observations of the Saltpêtrière on syringo-myelic dissociation in hysteria appeared, and most remarkable the patient in whom Charcot demonstrated this neurosic dissociation was studied by Gilles de la Tourette[90] from the point of view of the trophic, vaso-motor and oedematous troubles. Among these hysterics who "simulate syringo-myelic dissociation we find swollen, oedematous, cold, cyanosed hands which singularly resemble the "succulent hand," which Marinesco describes as in the true syringo-myelics.

Again hysteria has not so gross a physiology and associates in its manifestations, symptoms which often belong to organic lesions: so-called syringo-myelic dissociation and trophic and vaso-motor troubles.

In 1897 Marinesco[91] took up the question of

[88]Voir FRANCOIS FRANCK, art. Sueur in Dict. encycl. des sc. méd.

[89]PIERRET. *Acad. des sc.* 1882, et PUTNAM. Th. de Paris, 1882.

[90]GILLES DE LA TOURETTE et DUTIL. Contrib. à l'ét. des tr. troph. dans l'hyst. Atr. musc. et œd. *Nouv. Iconogr. de la Salpêtr.* 1889, p. 251.

[91]MARINESCO. De la main succulente. *Nouv. Iconogr. de la Salpêtr.* 1897, p. 84 et 202.

vaso-motor troubles in syringo-myelia àpropos to this oedematous hand which with P. Marie he called the "succulent" hand.

He recalled first many cases which had never been collected; then he added more recent cases: those of Massius (1890), Hoffman, Colemann and O'Carroll (1893), Dayot and Rennes (Louazelle 1890), with four personal observations. (Clinic of P. Marie and of Raymond).

He declares Pierret's opinion as to the location of the medullary center of the vaso-motor nerves not admissible, and accepts rather that of Remak[92] according to which this center is in the posterior grey horns; each segment of the cord containing its sensory and vaso-motor centers for the corresponding limb.

In fact, certain anatomists[93] admit today that the motor influence leaves the cord by the posterior roots and goes from there by the rami communicantes to the sympathetic ganglions.

It is permissible to think that these fibres which come from the posterior roots of the cord, arise in the posterior horns of the grey substance or in neighboring parts (Lenhossek and Cajal make them come from the base of the anterior horns "also from within and near the canal as well as from outside)."

Despite the obscurity which still persists on this question I believe we may conclude *clinically* that the *dissociation of sensation callled syringo-myelic is the syndrome of the posterior horns of the cord and that the vaso-motor and sudorific troubles*

---

[92]REMAK. *Berl. klin. Wochenschr.* 1889, n⁰ 3 (cit. Marinesco).

[93]Voir VAN GEHUCHTEN. Anat. du syst. nerv. de l'hom., 2e édit. 1897, p. 895, et POIRIER. Traité d'anat hum. 1884, t. III, p. 215.

(whén they are of medullary origin) *are the syn~ drome of the posterior and central grey matter* (base of the anterior horns). . ` .,

7. The Associated Syndrome of the Anterior Horns and of the Centro-posterior Grey Matter (the Syndrome of the entire Grey Matter): Muscular Atrophy, Dissocia- tion of Sensation called Syringo-myelic and Vaso-motor Troubles.

This syndrome, the anatomo-clinical and patho- logical physiology of which is explicitly contained' in paragraphs 4 and 6, is especially found in two classes of cases: syringo-myelias with amyotrophies and amyotrophies with vaso-motor troubles.

1. Syringo-myelia shows itself by amyotrophy and dissociation of sensation at a time when the lesion involves simultaneously the anterior and pos- terior portions of the grey matter: in these cases the amyotrophy has the characters which we have as- signed to amyotrophies of medullary origin.

Marinesco has insisted on the frequency of the . *preacher's hand* in these cases as described by Char- cot and Joffroy in hypertrophic cervical pachymen- ingitis and characterized by the predominance of atrophy in the regions of the ulnar and median nerves; further in these same patients the muscular. atrophy (of the Aran-Duchenne type) always has a very segmental distribution.[94]

2. In amyotrophics there is often noted hyper- thermia, the violet or livid hand (Vulpian). I have noted hyperthermia, and local sweats. Frommann,.

---

[94]"Voir le fait d'amyotr, en gant récem. publié par Crocq : (*Journ. de neurol.* de Bruxelles 1899).

Friedreich, Wunder, Leich, Leyden have also described exaggerated sweats in certain amyotrophic
cases.[95]

It seems useless to insist further on this point.

8. The Syndrome of the Lateral Half of the Cord: Crossed Hemiparaplegia.

A. In 1849, Brown-Sequard[96] described the
symptoms observed after a lesion of the half of
the cord: there is a crossed hemiparaplegia or
Brown-Sequard syndrome, characterized by *motor
paralysis and hyperaesthesia of the same side as the
lesion, and anaesthesia of the opposite side.*

More exactly we find: 1st. On the side corresponding to the lesion a motor paralysis, normal or
exaggeration of the various sensations, a zone of
anaesthesia in the upper portion of the part near the
lesion and often a zone of hyperaesthesia above, hyperthermia and paralysis of the sympathetic, arthropathies, muscular atrophy; 2nd. On the side
opposite the lesion there is preservation of voluntary
movement and of the muscular sense, anaesthesia
in all forms, sometimes a zone of hyperaesthesia
above the level of the lesion (the crossed sensory
troubles rising not quite so high as those on the side
of the lesion).

All authors have confirmed this *clinical description.* It has been completed by showing that often

[95]Voir notre Traité des mal. du syst., nerv. 4ᵉ éd (en col-
lab. avec Rauzier), 1894, t. I, p. 610.

[96]Les divers travaux de Brown-Séquard sur l'anesthésie
spinale croisée sont échelonnés depuis sa thèse en 1846
jusqu'en 1894. Les mémoires les plus importants sont ceux
de 1849 à la *Société de Biologie,* de 1863 dans le *Journal de
la physiologie,* de 1868 et 1894 dans les *Archives de physiologie.*

the anaesthesia is dissociated (of the syringo-myelic type) ; the same had been noted in certain observations collected at first by Brown-Sequard, especially those of Vigues. Sometimes there is only thermo-anaesthesia.

This unilateral dissociation is habitually crossed in relation to the lesions. More rarely it is direct, especially in the case of Dejerine and Sotttas.[97] In this latter case, the motor paralysis which is always direct (as in the classic cases) is on the same side as the anaesthesia; then we do not have the Brown-Sequard syndrome proper. However, these are cases which we ought not to overlook when studying the syndrome of the lateral half of the cord. We will take them up again àpropos to the pathological physiology.

The Brown-Sequard syndrome is clear and complete (as we shall describe it) when the lesion is exactly limited to the lateral half of the cord. We observe it in an attenuated and incomplete form when the lesion, more extended is more marked in one half than in the other. In this case the motor paralysis and anaesthesia are bilateral; only the paralysis is more marked on the side of the lesion and the anaesthesia (complete or dissociated) is more marked on the side opposite to the lesion.

Complete or incomplete, the Brown-Sequard syndrome is always a very good sign of the *medullary* location of the lesion. Outside of the cord *two* different lesions are necessary to produce this.

B. *The diseases capable of producing the Brown-Sequard syndrome* (complete or incomplete) are numerous. Any medullary lesion will produce

[97] Déjerine et Sottas, *Soc. de Biol.,* 23 juillet 1892. Obs. déjà citée.

it with the single condition that it involve exclusively a half of the cord or only one half of the cord more than the other.

We will note: traumatic lesions (fracture, luxation, hemorrhages, bullet wounds), vertebral arthritis, spinal meningitis, hemorrhagic foci, diffuse myelitis, syringo-myelia, tumors, various syphilitic localizations,[95] (syphilomas, myelitis, gummatous meningitis), etc.

C. The *pathological physiology* of this symptom still remains difficult and is much discussed despite the works which it has provoked for a half century.

It concerns us to study in extent only *crossed anaesthesia.* For, for the direct paralysis no one disputes the pathogeny which is very simple, the lesion involving the medullary motor fibres which have already crossed at the pyramids.

To explain the crossed anaesthesia, Brown-Sequard, in his first works, admitted that the medullary sensory fibres cross in the cord at every level of the cord from their entrance into the cord by the posterior roots. This explains the crossed anaesthesia; the hyperaesthesia above the level of the lesion is due to irritation of the surrounding part, as is also the hyperaesthesia of the same side as the lesion; the direct zone of anaesthesia at the level of the lesion is due to a change in the nerves at their entrance into the cord before their crossing.

Physiologically[99] Brown-Sequard shows anaes-

[95] Voir LAMY. La méningomyél. syphil. Paris. 1893.—
SOTTAS. Contrib. à l'étude anat., clin. des paral. spin. syphil. Paris 1894.—GILLES DE LA TOURETTE. Formes clin. et trait. des myél. syphil. Les Actualités medic. 1899.

[99] Voir pour tout ce paragr. DÉJERINE et THOMAS. Un cas d'hémiparapl. avec anesth. croisée. Syndr de Brown-Séquard suivi d'aut. *Arch. de physiol.* 1888, p. 594.

thesia of the hinder limbs after a double hemi-section, one in the dorsal region, and the other in the cervical region or after a longitudinal section separating into two lateral halves the lumbar enlargement of the cord, (Galien's experiment, remade by Fodera[100] 1823). The same experiment in the cervical region produced complete anaesthesia of both anterior limbs with preservation of sensation in the hinder limbs.

To this theory the physiologists have made and still make numerous objections and Vulpian [101] who was one of the first adversaries of Brown-Sequard's ideas, admitted that the hyperaesthesia (in a unilateral lesion of the cord) above and below the level of the lesion is produced by the local excitation and that the crossed anaesthesia is due to a depression correlative to the excitability of the analogous parts of the cord; the physiological balancing of the two halves of the cord.

These are some of the arguments of the adversaries to the theory of the intra-medullary intra-crossing of the sensory tracts.[102]

1. Among animals after unilateral section of the cord, if hyperæsthesia is the rule (Fodera, Vulpian) crossed anaesthesia is on the contrary "a phenomenon of very variable intensity in different animals and is not generally absolute."

2. Unilateral section has moreover less influence on the hinder limbs the further it is from the lumbar cord. (Vulpian).

3. In the experiment of Van Deen (unilateral

---

[100]FODERA (1823) et SCHOPS (1827) avaient entrevu le syndrome de Brown-Séquard.

[101]VULPIAN. Art. Moelle épin. (Physiol.) in Dict. encycl. des sc. méd. 1874, 2e série, t. VIII.

[102]Voir l'art. cité de DÉJERINE et THOMAS.

section toward the posterior portion of the dorsal cord and a complementary unilateral section in the cervical region) both hinder limbs retained sensation even if one unilateral section extended over the median line (dog, Vulpian; rabbit, Schiff).

, 4. The longitudinal section of the lumbar enlargement in the median line produces only a simple diminution of sensation (Oré).

5. More recently S. Mott[103] reached analogous conclusions for the monkey: painful and thermic sensations can be conducted by both sides of the cord, but touch sensations and those of pressure are principally conducted by the same side; there will be a considerable retardation in the transmission of the impressions received by the paralyzed side and some errors in localization. In résumé, in the monkey anaesthesia is principally direct.

6. Gotch and Horsley[104] have made researches as to the electric modifications which are produced in the various bundles of the dorsal cord when electric excitation is applied to the sciatic nerve or to the posterior roots of the lumbar plexus. In a way they measure by the intensity of this electric modification the quantity of nerve energy which passes into these bundles of the cord when the nerve in question is excited.

In the cat and monkey, the current in the cord is principally unilateral and of the same side in the proportion of 80 per cent and principally by the posterior column.[103]

[103]MOTT. Res. of hemis. of the spin. Cord in Monkeys, *Philos. Transact.* 1891, Cit. Déjerine et Thomas.—Voir aussi la discussion de ces expér. par Marinesco in *Sem. méd.* 1896, p. 263.,

[104]GOTCH et HORSLEY. On the mammalian nerv. syst., its Funct. and their Localis. determ. by an electr. Meth. *Philosoph. Transact.* 1891. Cit. Déjerine et Thomas.

Many months after the hemi-section of the cord the electric modification above the hemi-section is three times less strong than if the excitation is applied to the opposite side.

In résumé Déjerine and Thomas grouping the facts of experimental physiology[106] conclude that the intra-crossing of the sensory tracts is not complete in the cord: undoubtedly a partial intra-crossing occurs; that in certain animals the intra-crossed fibres are more numerous than the direct; in other animals (the monkey and the cat) the contrary is a fact.

Impressed by all these arguments Brown-Sequard[107] himself toward the close of his life abandoned his first theory. He recalls that a puncture of the posterior cord of one side can produce the syndrome; that after first a unilateral section in the cervical cord with consecutive hemianaesthesia, if a second dorsal unilateral section were made hyperaesthesia replaces the hemianaesthesia and vice versa; that hemianaesthesia consecutive to a unilateral section of the cord may disappear after a stretching of the sciatic nerve on the side of the anaesthesia. And then according to him the anaesthesia in his syndrome becomes a matter of inhibition and the hyperaesthesia a matter of over-activity.

---

[106]Il seriat utile d'établir que, chez ces animaux, l'excitation électrique est comparable, pour les voies de conduction intramédullaire, à l'excitation venue des organes sensitifs périphériques.

[106]Voir aussi Bottazzi. Sur l'hémis. de la moelle épin. *Riv. sperim. di fren.* 1896, t. XXI et*Revue neurol,* 1896, p. 372.

[107]Brown-Séquard. Rem. à propos. des rech. du Dr Mott sur les effets de la sect. d'une moitié latér. de la moelle épin. *Arch. de physiol.* 1894.

This opinion so plainly settled has separated the clinicians from the physiologists. And so Brissaud[108] has continued to defend and to represent in a scheme the intra-medullary dissociation of the sensory tracts, Raymond[109] appeared disposed to take up Brown-Sequard's last opinion, and Déjerine and Thomas, recognizing that it is formulated in "very vague terms" are tempted however "to take into consideration the last opinion of Brown-Sequard, already moreover sustained by Vulpian." And they add "Whatever it is, it is impossible to put Brown-Sequard's theory on a solid basis, especially when theories[110] are preferred to facts."[111]

It seems to me that the question can be presented in a less discouraging manner. Without lessening the value of the conclusions accumulated by experimental physiology it may be remarked that all these experiments establish a single thing: the mode of intra-medullary transmission of sensory impressions in animals.

But in man things may take place differently:

[108]BRISSAUD. Hémiparapl. spin avec hémianesth. croisée, syndr. de Brown-Séquard in Leç. sur les mal, nerv. 1895, p. 246, (leç du 15 déc. 1893).—Le double syndr. de Brown-Séquard dans la syph. spin *Progrés méd.* 1897 nos 29 et 51. Voir aussi LONDE. Double syndr. de Brown-Séquard dans le mal de Pott. *Revue neurol.* 1898, p. 356.

[109]RAYMOND. Synd. de Brown Séquard d'orig. probabl. syringomyél. (Leç. du 28 juin 1895) in Clin. des mal. du syst. nerv. 1896, t. I, p. 315 et sur un cas d'hemis, traumat. de la moelle (syndr. de Brown-Séquard). Leç. du 20 nov. 1896 in Clin. des mal. du syst. nerv. 1898, t. III, p. 508.

[110]Le schéma qui est un moyen indispensable d'enseignement ne me paraît pas si dangereux tant qu'il reste ce qu'il devrait toujours être: un résumé et une expression synthétique, toujours revisables, des faits observés.

[111]Voir aussi les idées de Déjerine dans la thèse (Paris 1899) de Zong sur les voies centrales de la sensibilité générale.

the specialization of functions in the nervous system is always increasing in proportion as you go up the animal scale. If this principle, which I believe indisputable, is admitted, it must be recognized that the *anatomo-clinical method* is the only one which can decide whether one should or should not apply the conclusions of experimental physiology to man.

That is to say that the *clinical facts* if they are pretty numerous and well observed, have their value by the side of and in the face of *physiological facts.*

To 24 observations collected by Brown-Sequard in his memoir of 1863 a great many new facts have been added, many of which are recent and with autopsy and all establish the reality of Brown-Sequard's syndrome; that is to say prove that in a unilateral lesion of the cord there is a direct motor paralysis and crossed anaesthesia.

Then in conclusion, it seems to me irrefutable that *things take place from this point of view differently in man and animals.*

If Brown-Sequard's syndrome is admitted as a clinical law (and this appears certain) the intramedullary crossing of the sensory fibres in man must be admitted. We come back then to Brown-Sequard's first theory which alone explains or at least expresses the clinical fact.

Anaesthesia is direct in a limited region where the sensory fibres are injured at their entrance into the cord before their intra-crossing; it is, on the contrary, crossed in the more extended region where the sensory fibres are injured in their intra-crossing. Hyperaesthesia develops by irritation of the neighboring parts in limited regions where the sensory fibres pass into the cord on the side of the lesion.

By combining this with what has already been said in the pathological physiology of the dissociation called syringo-myelic, it can be understood how crossed dissociation occurs in certain Brown-Sequard cases.

D. The *differential diagnosis is short.* The Brown-Sequard syndrome complete or incomplete is in a word characteristic and corresponds always to the medullary site of the lesion.

A single case should be distinguished, that of a *double lesion.* The syndrome is pathognomonic of a medullary origin only when the direct paralysis and crossed anaesthesia can be explained by a single lesion.

If there is a double lesion (extra-medullary, root, or neuritic) the distribution of the affection will generally follow the nerve territories instead ot being segmental; further the two lesions will generally have appeared at different times and each one will have its peculiar independent symptomatology.

II. DIAGNOSIS OF THE HEIGHT OF THE LOCATION OF MEDULLARY LESIONS.

1. General Principles as to Diagnosis of the Height of Lesions.

In the preceding chapter, studying the semeiology of the various systems of the cord we have endeavored to make the diagnosis of the *breadth* of the location of the medullary lesion.

It remains now to indicate the elements of a diagnosis as to the *height* of the lesion.

Being given a lesion of the cord, the following elements useful in making this diagnosis may be derived:

1st. The *external signs* of the initial lesion: fracture, luxation, deviation, gibbosity. To give to these signs (when one finds them) their full semeiological value as to the medullary location, it is necessary to call attention to the correspondence between the spinous processes (the part most accessible to clinical exploration) and the vertebral bodies and in consequence to the roots of the various spinal nerves.

Here are the indications given by Chipault[112] on the relation of the spinal processes to the origin of the spinal roots: "In the adult, in the cervical region add *one* to the number of a determined process to give the number of the root which rises at this

---

[112]CHIPAULT. Sur les rapp. des apoph. épin. avec la moelle, les rac. médull. et les mén. Th. de Paris 1894—Cit. RAYMOND. Clin. des mal. du syst. nerv. 1896, t. I, p. 275.

level; in the upper dorsal region, add *two;* from the 6th to the 11th dorsal process add *three;* in the lower part from the 11th and the subjacent inter-spinous space they correspond to the last three pairs of lumbar nerves, the 12th dorsal process and the subjacent space correspond to, the sacral pairs.."[112]

2. The *motor paralysis and anaesthesia* fur-nish valuable information by their distribution and especially by their upper limit.

When the lesion affects an entire section of the cord the paralysis and anaesthesia involve the part below the lesion particularly, but if the lesion is par-tial only the region whose innervation depends on the zone destroyed is affected.

3rd. A more disputable and today still more dis-cussed question is that of the semeiological value of the *state of the reflexes.*

At first sight the matter appeared very simple to the clinicians who applied the classic formula of the physiologists: the reflex power of the cord is increased in the parts separated from the higher centers by a section of the cord; the brain normally exercises an inhibitory action over the reflex power of the cord; when a section or a lesion hinders the passage of this inhibitory action, the reflexes are ex-aggerated below the section or the lesion. On the other hand, when the lesion involves a zone of the cord the reflexes which have their center exactly in this zone will naturally be lost.

From this arises this clinical rule, which was of

[112]On trouvera à la page 657 du Traité d'anat. hum. de Testut. 3e édit. 1897, t. II, un tableau complet des "origines spinales des nerfs rachidiens rapportées aux apophyses épin-euses."

great assistance in the diagnosis of the height of the medullary lesion: *when a lesion affects a section of the cord the reflexes whose centers are below the lesion are exaggerated, and the reflexes whose centers are at the level of the lesion are lost.* This is the classic law applied in transverse myelitis and in compression of the cord.

But there are facts which seem to weaken this law. In 1890 Bastian[114] published four cases in which the cervical or dorsal cord was entirely destroyed and where the reflexes were abolished, the bladder and rectum alone remaining intact. From this the attention was directed to the flaccid paraplegia of certain cases of transverse myelitis and of certain compressions of the cord and other cases have been published, first by Bowlby, Rooth, Babinski[115] and ourselves.[116] Vulpian[117] had mentioned the disappearance of the reflexes in certain cases of transverse myelitis where the lesion was deep, and Kadner (1876), Weiss (1878), Kahler and Pick (1880), Schwartz (1882), Thorburn[118] (1887-1888) have published analogous cases.

More recently the observations of Jackson (1892), Bruns (1893), Gehrardt (1894), Hitzig (1894), Egger (1895), Hoche (1896), Habel (1896), and the important memoirs of van Ge-

---

[114]BASTIAN. *Brit. med. Journ.*. 1890, p. 480. Anal. in *Rev. des sc. méd.*, t. XXXVI, p. 520—Premiers trav. du méme sur ce sujet en 1882 et 1886.

[115]BABINSKI. Parapl. flasque par compress. de la moelle. *Arch. de méd. expérim.* 1891, p. 228.

[116]Mal de Pott et parapl. flasque anesthés. (leç. de 1893) in *Leç. de clin. méd.* 1896, 2e série, p. 372.

[117]Cit. BRISSAUD.

[118]Cit. VAN GEHUCHTEN.

huchten,[119] Marinesco[120] and Brissaud[121] have appeared.

From all these documents it must be concluded that the ancient classic law of the reflexes is no longer true, and that in a certain number of cases there is an abolition of reflexes whose medullary centers are below the level of the lesion. It is less easy when the theory of these facts is attempted.

Bastian has taken the opposite of the classic theory: in the normal state the action of the higher centers is necessary to the medullary reflex; when this dynamogenic action is suppressed by a section or complete lesion all the reflexes below are abolished; the reflexes are preserved or exaggerated only if the transverse destruction of the cord is partial. This theory explains very well the new facts of flaccid paraplegia, but fails by taking no account of spastic paraplegia; though these cases (as in the observations of physiologists in animals) exist and perhaps are in the majority. To meet Marinesco's opinion Brissaud has shown that there are cases of transverse and complete lesion, verified by autopsy, with preservation or exaggeration of reflexes. Such cases are those of Gehrardt, Senator and those which Brissaud communicated to the Congress at Angers. Then, the theory of Bastian is also as im-

---

[119]VAN*t* GEHUCHTEN. Le mécan. des mouv. réfl. Un cas de compress. de la moelle dors. avec abolit. des réfl. *Journ de neurol.* 1897, p. 262, 282, 302 et 322; et Etat des réfl. et anat. pathol. de la moelle lombo-sacrée dans les cas de parapl. flasque *qus* à une lés. de la moelle cervico-dors. *Ibid.* 5 juin. 1898.

[120]MARINESCO. Sur les parapl. flasques par compression de la moelle. *Sem. méd.* 1898, p. 150.

[121]BRISSAUD. La parapl. flaccide par compress. *Revue neurol.* 1898, p. 350; et Congrès d'Angers, août 1898, (*Ibid.* p. 589).

5

possible to sustain as the old classic theory : both are too exclusive and neither explains all the facts.

Van Gehuchten's theory is derived from Bastian's only the Louvain professor adds the notion of a double cortico-spinal path maintaining the medullary tonus by the action of higher centers : a direct inhibitory path, an indirect (by the medulla) exciting : a lesion of the pyramidal fibres exaggerates the reflexes ; a complete lesion of the cord suppresses all the reflexes. This theory is liable to the same objections as Bastian's.

A definite theory does not appear to have been found. However, Brissaud showed in a great many observations of cases of flaccid paraplegia an alteration of the nerves or of the medullary cells below the lesion, that is at the level of the paralysis.

If this view were accepted, the old classic theory might persist.

When the lesion is cervico-dorsal the reflexes of the lumbar region will be exaggerated so long as there is no secondary lesion of the nerves or lumbar cells. When this secondary lesion exists the reflexes having their centers in this region will naturally be abolished.

This is what I expressed theoretically in 1893 when I said,[122] "The gibbosity occupying the cervical or dorsal region, the paraplegia will be dorsolumbar in its aspect ;" and by its secondary lesion, adds Brissaud.

Pierret[128] accepted this view and declared "That the lack of secondary contracture could be attributed to a peripheral neuritis."

---

[122]Leç. citée sur le mal de Pott et la parapl. flasque anesth. p. 394.

[128]PIERRET. Congrès d'Angers, août 1898, p. 583. *Revue neurol.* 1898, p. 583.

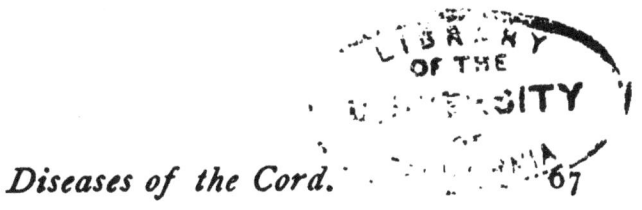
Whatever these disputable theories may be let us remember that the reflexes are not always exaggerated nor preserved below the lesion. Then, when they are exaggerated, this can be used in the diagnosis of the height of the medullary lesion; but when they are lost they cannot always be utilized in this direction.

4. These things said, we will study successively the semeiology of the conus medullaris, sacral, lumbar, dorsal, brachial and cervical cords.

The syndromes may be divided into two groups: the syndromes of the root distribution and syndromes of the metameric or segmental distribution.[124]

It is known that for the same part of the cord the distribution of symptoms is not identical and the two orders of symptoms are not superimposed. The root distribution is observed when the medullary lesion involves the zones of entrance of the roots, and the segmental when the lesion affects the centers themselves of the cord. Moreover, physiological segmentalization can be naturally observed only in the lumbar cord and above. From this it results that we will have only a single syndrome (radiculo-segmental) for the conus and sacral cord while for the portions above we will separately describe the root and segmental syndrome.

[124]"Segmentaire" veut dire ici symptôme se distribuant "par segment de membre."

2. The Root-Segmental Syndrome of the Conus Medullaris.[125]

The anatomists[126] limit the terminal conus to the level between the last sacral and the first coccygeal pair.

The clinicians include the last two (Raymond) or three (Müller) sacral pairs.

In favor of this last limitation Müller has given arguments derived from the histological structure: the anterior roots there become scarce and the pyramidal tracts disappear.

Let us then adopt this definition of the conus: the most inferior portion of the cord[127] from which arise the last three sacral pairs (3rd, 4th and 5th) and the coccygeal nerves.

To philosophically retain the distribution of the nerves of the conus it is necessary to place man on four feet; then the sensory region of these nerves is the most posterior surface of the body (sacrum, coccyx, anus and perinaeum), as to the motor distri-

---

[125]Voir nos leç. sur les paral. nucléaires des nerfs sacrés in *Leç. de clin. méd.* 1898, 3e série, p. 249; DUFOUR. Contrib. à l'étude des nerfs de la queue de cheval et du cône terminal. Th. d Paris 1896; RAYMOND. Sur les lés. de la queue de cheval (14 et 21 déc. 1894)et hématomyélie du cône terminal (24 mai 1895) in *Leç. sur les mal du syst. nerv.* 1896, t. I, p. 252 à 314; MULLER. Untersuch. üb d. Anat. u Pathol. d. untersten Rückenmarksabsch. (Clin. méd. du prof. Strumpell à Erlangen) D. Zeitschr. f. Nervenh., t. XIV, p. 1, (22 déc. 1898). Bonne bibliogr. p. 88 et Schméas intéressants, pl. I et II.

[126]CHARPY in Traité d'anat. hum. de Poirier.

[127]On pourrait rattacher au cône médullaire le filum terminal. Mais si, pour l'anatomiste, ce filum représente une moelle coccygienne ou caudale, plus ou moins rudimentaire, on peut dire qu'il n'existe pas pour le clinicien; à ce niveau il n'y a plus pour lui que la queue de cheval.

bution, it is to the muscles also of the analogous regions.

This settled or accepted, the following (according to Müller) are the motor functions and the sensory region of this segment of the cord which is situated at the level of the second lumbar vertebra:

CONUS MEDULLARIS.

	Motor Function.	Sensibility.
3rd. Sacral segment.	Center of ejaculation Ischio- and bulbocavernous.	Skin of penis and part of scrotum; Urethral mucous membrane.
4th. Sacral segment	Vesical centers Detrusor vesicae	Skin of perinaeum and sacrum.
5th. Sacral segment and Coccygeal segment.	External sphincter of anus; relaxer of sphincter.	Skin of coccyx and of anus.

From this table the syndrome of the conus can be deduced :[128] It is essentially made up of troubles of micturition and of defecation (obstinate constipation or relaxation on the part of the bowels; retention or incontinence of urine), absence of erection and anaesthesia of the penis, urethra, scrotum, perinaeum, anus, coccyx and of the sacrum. This picture is more or less complete according to the case.

The morbid causes which can determine this syndrome are: Traumatism (a blow in this location, a shot producing a fracture, a hemorrhage), tumors (Raymond names lipomas, myomas, sarcomas, gliomas, cavernous lymphangiomas, medullary cancers), syphilis, tuberculosis, meningeal hemorrhages, etc. (Duwour).

As to the differential diagnosis that must be made from lesions of the cauda equina. This will be best considered after the study of the syndrome of the sacral cord.

3. The Root-Segmental Syndrome of the Sacral Cord.[129]

The sacral cord as to location has nothing to do with the sacrum. It corresponds to the first lumbar vertebra and gives rise to the first sacral and the last lumbar pairs. Following is its motor and sensory distribution according to Müller :[130]

---

[128]RAYMOND cite les faits de Lachmann (1882), Kirchoff (1884), et Oppenheim (1889), Lubovitch (1894, *Revue neurol*, 1895, p. 20) et PETERSON (1895, *Revue neurol*, 1895, p. 412) ont contesté qu'une lésion limitée au seul cône ait pu déterminer la paralysie complète de la vessie et du rectum.

[129]Voir les travaux déjà cités pour le syndrome du cône médullaire.

[130]SANO (*Journ. de neurol.* 1897, p. 277) place le centre des muscles du pied et des muscles de la jambe dans une

SACRAL CORD.

	Motor Functions.	Sensibility.
5th. Lumbar segment.	Abductors: Middle and small glutei; tensor of the fascia lata semi-membranous, femoral biceps.	External part of the thigh.
1st. Sacral segment.	Rotators outwards: Pyriformis, internal obturator, the gemelli, gluteus maximus.	Posterior part of thigh and leg.
2d. Sacral segment.	Large muscles of the calf: gastrocnemius, soleus, tibial, peroneal muscles. Center of erection.	External part of leg and of foot. Sensibility of the bladder and of the superior portions of the large intestine.

colonne de la 4e sacrée à le 5e lombaire; le centre des fessiers, de la 2e sacrée à la 5e lombaire; celui du quadriceps fémoral. de la 4e a la 2e lombaire; celui des muscles abdominaux, de la 1re lombaire et au-dessus...Ces résultats ont été discutés à la Soc. belge de neurol.—Van Gehuchten et de Buck (*Revue neurol.* 1898, p. 510) concluent de leurs recherches: "1o les noyaux d'innervation des muscles de la jambe et du pied occupent la partie postérieure des cornes antérieures de la moelle et s'étendent depuis la partie supér-

To synthetize this table, placing man, as for the
conus on four feet and putting the lower limb in its
primitive position (turned outward at an angle of
90 degrees from the positive position) the internal
surface and the large toe forward, we find the sacral
cord presides over the sensibility of the posterior
and external surfaces of the lower limbs corresponding to a long posterior surface of this lower
limb (including the sole of the feet) ; the muscles
(rotation outward, flexor of leg and extension of
foot) are also those of the posterior surface.

Following are the various partial paralyses, the
superposition of which constitutes the total motor
syndrome of the sacral cord: pes valgus, (long peroneus), pes varus (short peroneus), impossibility
of flexing the foot on adduction (anterior tibial
group), impossibility of flexing the foot on abduction (long extensor of the toes), foot drop, steppage
(whole of the external popliteus), equilibrium in
walking[131] but not static coordination; impossibility
of extending the foot, of flexing and turning the
toes laterally (internal popliteus), impossibility of
flexing the leg on the thigh (semi-tendinous, semi-
membranous, biceps) ; abduction of the thigh im-

ieure du 5e segment lombaire jusque vers l'extrémité inférieure du 4e segment sacré; 2º il existe deux grands noyaux
d'innervation de ce segment du membre inférieur; un premier noyau très grand, comportant probablement plusieurs
subdivisions, s'etend de l'extrémité supérieure du 5e segment
lombaire jusqu'à la partie inférieure du 3e segment sacré;
und second noyau, également assez volumineux, surtout vers
son milieu, mais semblant unique, commence, en arrière du
premier, à partir du 2e segment sacré et s'étend jusque vers·
l'extrémité inférieure du 4e segment sacré."

[131]Voir mes leç. sur un cas de pseudo-tabes post-infectieux
(par. symétr. postérysipel. du tibial antér.), in Leç. de Clin.
méd. 1896; 2e série, p. 245.

possible, rotation difficult, difficulty in climbing stairs (glutei).

For the sensibility the anaesthesia includes: in front, the upper part of the foot and the external part of the leg; behind, all, except the internal part of the leg and the lower half of the thigh.

To make the sensory syndrome of the sacral cord complete it is necessary to add to this distribution the anaesthesia of the conus medullaris, the topography of which we described above.

The pains, when they exist, will be in the domain of the sciatic.

But, to appreciate the semeiological value of a pain from the point of its location it must be recalled that it is there a phenomenon of excitation and not of destruction, and that in consequence the seat of a pain often indicates a point near the diseased region rather than the altered part itself.

To complete the syndrome it is necessary to know the reflexes which have their centers below or at the level of the sacral cord.

We already know that in the conus are the anovesical and genital reflexes. In lesions of the sacral cord these reflexes will often be exaggerated (retention and priapism) at other times diminished or abolished (incontinence, impotence).

In the sacral cord are the plantar and tendon Achilles reflexes.

The plantar reflex is a cutaneous reflex which we studied above àpropos to its qualitative alteration described by Babinski. In the normal state it responds, according to Babinski,[132] by a general flexion of the thigh on the pelvis, of the leg on the thigh, and of the toes on the metatarsals. Bris-

---

[132]BABINSKI. *Soc. de biol.* 22 févr. 1896. Cit. Ganault.

saud[133] studied the slightest manifestations of this reflex and demonstrated them in the isolated contraction of the adductors (adduction of the point of the foot) and especially in the isolated contraction of the tensor of the fascia lata.

Ganault[134] confirmed Brissaud's conclusions and added for other cases the contraction of some special foot muscles.

The center of this plantar reflex is in the sacral cord: center of the 2nd and 3rd sacral roots when it acts only by movements limited to the toes, 5th lumbar segment when there is contraction of the tensor of the fascia lata, the whole of the sacral cord when plantar excitation causes all the reflex flexion of the lower limb.

This is the first reflex which will be abolished in destructive lesions of the sacral cord.

The reflex of the tendon Achilles has its center in the first sacral segment (Gowers).[185] It also will be abolished in destructive lesions of the sacral cord.

With the whole of these considerations of the nervous troubles, the anaesthesia and the state of the reflexes we have an entire syndrome of the sacral cord.

The *differential diagnosis* of this syndrome of the sacral cord is to distinguish it from that of the cauda equina. It is at times difficult, at least when it affects the lower portion of the cauda equina which corresponds to the sacrum; for in its upper part it includes the lumbar and sacral roots together, and the diagnosis must be made rather from the

---

[133]BRISSAUD. Le réfl. du fascia lata. *Gaz. hebdom.* 1896, p. 253.
[134]GANAULT. Contr. à l'étude de quelques réfl. dans l'hémipl. de cause organ. 1898, p. 86.
[135]GOWERS. Cit. Sternberg.

syndrome of the lumbar cord. But in its lower part the cauda equina includes only the sacral roots, wherefore its alteration will give the same symptomatology as the alteration of the centers of these roots. For the sacral nerves it is a question (which occurs for all nerves) of differential diagnosis between nuclear and root paralyses (or neuritic) ; for the root lesions are intra-spinal neuritic lesions.

Nothing proves better the difficulty of this diagnosis than the case in which Erb[186] claimed a lesion of the cauda equina, and at autopsy seven years later Schultze[187] established a lesion of the cord. Bechterew[188] also thinks "That it is impossible to know whether a lesion involves the cord or the roots to which the medullary segment gives rise."

I however believe this diagnosis possible, at least in certain cases.[189]

Nothing can be deduced from the anaesthesias, paralyses or amyotrophies. They are the same for the nerves whatever may be the height of the lesion. The reaction of degeneration, the symmetry of the affection, the cause, the onset and evolution of the disease can do nothing to make the diagnosis clearer.

On the contrary, I believe the consideration of the following four orders of symptoms are useful:

1. The objective and often external signs which indicate the location and height of the lesion: spon-

---

[186]ERB.. Ueb. Spinallahm (poliomyél. antér. acuta) bei Erwachs. u. üb. verw. spin. Erkrank. *Arch. f. Psych.* 1875, t. V, p. 758. Obs. VI.

[187]SCHULTZE. Z. différent. Diagn. d. Verletz. d. Cauda eq. und d. Lendenanschw. *D. Zeitschr. f. Nervenh.* 1894, t. V, p. 147.

[188]BECHTEREW. Cit. DUFOUR, *loc. cit.*, p. 62.

[189]Voir mes leç. citées sur les paral. nucléaires des nerfs sacrés, p. 269.

taneous or provoked pains,[140] gibbosity, displacement. These signs, often very useful, are not of absolute value. An intra-spinal post-traumatic hemorrhage may not show and especially remain at the very place of the traumatism. Dufour has indeed remarked that the dura mater in the cauda equina protects the lumbar roots more than the sacral roots.[141]

2. The dissociation of certain reflexes when it exists: I speak of the abolition of certain reflexes and the exaggeration of others below.[142]

3. The syndrome of Brown-Sequard when it exists (which is very frequent) anaesthesia more marked on one side, paralysis more marked on the other, proving a medullary origin.

4. The dissociation called syringo-myelic, when found, is a good sign of medullary origin despite the cases which we have cited above and in which the said dissociation would have been produced by a neuritis.

4. The Root Syndrome of the Lumbar Cord.

The *lumbar cord* which corresponds to the first four lumbar roots is at the level of the bodies of the 10th, 11th and 12th dorsal vertebrae. Following, still according to Müller, are its sensory and motor regions:

[140]Les douleurs sont plus fréquentes dans les lésions de la queue de cheval que dans lésions de la moelle sacrée.

[141]Le sac dural se termine au niveau de la 2e vertèbre sacrée et à partir de ce point les racines sacrées (portion inférieure de la queue de cheval) sont plus exposées, notamment aux hémorragies extra-durales, qui ne pouvant pas filtrer à travers le duremère se collecteraient plus bas dans le canal sacré.

[142]Nous retrouverons mieux ce signe à propos de la moelle lombaire.

LUMBAR CORD.

	Motor Functions.	Sensibility.
1st. Lumbar segment.	Lower part of the muscles of the abdomen.	Skin of lower part of abdomen.
2d. Lumbar segment.	Psoas, iliacus internus, cremaster.	Testicle, spermatic cord. Lower part of the hips. Mons Veneris.
3d. Lumbar segment.	Sartorius, pectineus, adductors.	Anterior and internal part of hip.
4th. Lumbar segment.	Quadriceps femoris, internal rectus, external obturator.	Anterior and internal part of thigh. Narrow band on the internal part of the leg to the internal border of the foot.

On placing man, as for the sacral cord, on four feet and the lower limbs in the primitive position, we see that the lumbar cord presides over the sensibility of the internal and anterior surfaces, that is to say of the anterior segment of the lower limb and over the mobility of the same regions (adductors, rotators within, extensors of the leg).

To sum up the distribution of these three segments of the cord, we see that man being in the position given above, the three segments of the cord which are the one in front of the other (the conus behind, the sacral cord in the middle, the lumbar cord in front) respectively innervate three segments placed also one before the other: the conus innervates the caudal segments (or its location), the sacral cord innervates the postero-external part of the lower extremity, and the lumbar cord innervates the antero-internal part of the lower extremity.

These synthetic views should aid in fixing in the memory this sensori-motor distribution of the three first (lower) segments of the cord.

In every case from the above table it should be easy to deduce the syndrome produced by an entire section of the lumbar cord.

In a word, there is complete paraplegia with anaesthesia up to the lower part of the abdomen, sphincter troubles and often bedsores on the sacrum, amyotrophy when present is limited to the muscles which the table gives as directly innervated by the lumbar cord. The muscles depending upon the sacral cord are paralyzed because their communication with the brain is interrupted, but they escape atrophy because their medullary center is not changed. When there is pain it is generally at the limits of the lesion; above under the form of ileolumbar neuralgia, below under the form of crural or sacral neuralgias.   .

As to the reflexes we must always distinguish between those whose center is below the lesion and those whose center is at the level of the lesion.

In the first group we have the sphincter reflexes,

the tonus of the muscles innervated by the sacral cord, the plantar reflex and the tendon Achilles reflex. When (which is most frequent) the reflexes are exaggerated there is a spastic paraplegia (at least in the sacral region), that is to say the contractures produce pes equinus; there is clonus and retention of urine and faeces.

Of the reflexes having their center in the lumbar cord itself the patellar tendon is the principal one. The roots of the lumbar plexus make up the arc.[143] The roots, the integrity of which seems necessary for the maintenance of the patellar reflex, are in the rabbit (Tschiriev) the 6th lumbar, in the dog (Westphal) 5th, 6th, 7th lumbar, in man (Gowers) 2nd, 3rd and 4th lumbar. So that the center of this reflex is in the lumbar cord. Then, the abolition of the patellar tendon should make a part of the syndrome of the lumbar cord.

As, in another place, we have seen that with a lesion in this location the reflex of the tendon Achilles may be exaggerated, we would have this strange dissociation an example of which we have published,[144] on one part abolition of the patellar tendon reflex,[145] and on the other in the same limb clonus.

Another reflex (a skin reflex) also has its center

[143]Voir STERNBERG. D. Sehnenrefl. u. ihre Bedeut f. d. Pathol. d. Nervensyst. 1893, p. 34.

[144]Leç. de clin. méd. 1898, 3e série, p. 252.

[145]On pourrait parler ici du réflexe contralatéral des adducteurs de P. Marie. Ce réflexe (voir la Th. citée de Ganault, p. 56) es provoqué par la percussion du tendon rotulien et peut persister alors que le réflexe rotulien ordinaire est aboli, mais la condition pathogénique intra-médullaire de ce phénomène n'est pas encore assez nettment établie pour que nous nous y étendions ici. Il en est de même du réflexe rotulien paradoxal, c'est-à-dire des sujets chez lesquels la percussion du tendon rotulien fait fléchir la jambe au lieu de la soulever.

in the lumbar cord, the cremasteric.[146]   The quick elevation of the testicle is produced by friction or brusque pressure on the skin of the supero-internal part of the thigh or better at the level of the ring of the third adductor.   In the female there would be an analogous reflex from the groin, a contraction of the most inferior fibres of the abdominal wall.   The center of this reflex is in the lumbar cord (1st and 2nd segments).

Then the patellar and cremasteric reflexes will be abolished in lesions of the lumbar cord and this abolition is contrasted with the maintenance or even exaggeration of the tendon Achilles and plantar reflexes.

The *differential diagnosis* is easier than for the sacral cord for lesions of the cauda equina cannot produce complete paraplegia with sphincter troubles. The question will arise only when the syndrome of the lumbar cord is reduced and incomplete.

On the same principle given above for the sacral cord we may base the following: the objective and external signs will indicate the height of the lesion in the spinal column, the dissociation of the reflexes (very important), the Brown-Sequard syndrome (if it were only indicated) and the so-called syringo-myelic dissociation of sensibilities.

5. The Segmental Syndrome of the Lumbo-Sacral Cord.

All we have said in paragraphs 2, 3 and 4 is based on the distribution of the spinal roots.   It is therefore the root semeiology.   These are the *root syndromes of the lumbo-sacral cord.*   By the side of

[146]Voir GANAULT. Th. citée, p. 3.

these it is necessary to look at also the *segmental syndromes of the lumbo-sacral cord.*

The characteristic symptom is that it is *segmental:* there is often here a complete or dissociated anaesthesia; it does not correspond then either to a nerve or root distribution, but to a *segment of a limb;* its upper limit is a circular line perpendicular to the axis of the limb.

When this segmental anaesthesia is limited to the foot, it may be confounded with a nerve or root anaesthesia (sacral plexus); when it extends over all the lower extremity it may be confounded with an anaesthesia from all the lumbo-sacral plexus. But when it involves, for example, all the lower part of the limbs and is limited above by a circular line at the lower third of the thigh, neither the nerve nor root distribution can be invoked, the segmental distribution must be admitted.

Debove and Parmentier[147] have observed some cases of syringo-myelia in which thermo-analgesia or thermo-anaesthesia were thus disposed as stockings.

Chipault[148] has described anaesthesia from Pott's disease in the shape of boots, in that of long stockings (up to the middle of the thigh) in that of drawers (up to the umbilicus), and a hyperaesthesia up to the middle of the thigh in a medullary disturbance.

Thus, although the segmentalization may be less distinct and less frequent in the lower limbs than in

[147]DEBOVE, in Leç. du mardi de Charcot, t. II, p. 506.— PARMENTIER. Nouv. Iconogr. de la Salpêtr. 1890, p. 219. Cit. de BRISSAUD, in Leç. citées sur les mal, nerv., p. 220.

[148]CHIPAULT. La topogr. de l'anesth, pottique. *Revue neurol.* 1896, p. 293, et Quelques types clin. nouv. de lés. radicul. et médull. *Presse méd.* 1896, p. 85.

the upper, the clinician must admit it. The segmental syndromes are recognized by their segmental distribution: the seat of the lesion is in these cases in a section of the lumbo-sacral cord as much higher as the segment of the limb attacked is itself higher.

6.   The Root Syndrome of the Dorsal Cord.

The *dorsal cord,* which extends from the 2nd to the 9th dorsal vertebræ, corresponds to the origin of the dorsal pairs of nerves from the 2nd to the 12th.

Each of these nerves innervates, from a sensory point of view, a band of the trunk. The upper band (2nd dorsal) extends from the upper part of the internal surface of the arms. From the 3rd to the 4th dorsal nerves the zones are above the breast, which is included in the zone of the 5th; from this to the navel is included in the zone of the 10th.[149] The zone of the 12th is confined to the upper zone of the lumbar cord.[150]

These zones, horizontal to the upper part of the trunk, become more and more oblique from above downward and from behind forward in proportion as you descend.

Each zone is innervated principally by the corresponding pair of nerves and accessorily by the pairs immediately above and below (Sherrington). It results, that when a root is cut, the anaesthesia is not complete in the corresponding region and that in a given anaesthesia it is necessary to look for the

[149]Un fait de MACKINTOSH (*The Brit. med. Journ.* 1898, p. 478. *Revue neurol.* 1898, p. 203) montre que la zone sensitive de la dixième racine dorsale n'atteint pas l'ombilic.

[150]Voir la distribution d'après THORBURN in MARINESCO, *Sem. méd.* 1896, p. 259. (Travaux de Sherrington, Horsley....).

lesion four inches above the line of anaesthesia (Horsley).

For the motor distribution see the following table according to Testut:

INTERCOSTAL AND EXT-INTERCOSTAL.		*Muscles.*	
DORSAL. *Pairs of Nerves.*		*Muscles.*	DORSAL. *Pairs of Nerves.*
2nd., 3rd., 4th.		Serratus posticus superior.	
5th., and 6th.		Triangularis sterni, ext. oblique, rectus abdominis, (9th., 10th., 11th., 12th.)	
		Int. oblique, transversalis.	7th. and 8th.
9th., 10th., and 11th.		Serratus posticus inferior. Large muscles of the abdomen.	
		Pyramidalis.	12th.

From the anatomo-physiological facts the syndrome of the dorsal cord may be easily deduced: Paralysis and anæsthesia, in a complete lesion of the cord, in the whole region, motor and sensory, located below the lesion. Beside the bedsores, which are as in the lumbo-sacral cord, there are other important trophic disturbances: there may. be a herpes of the trunk, which then outlines the distribution of the roots.

For the reflexes we will have loss and more often exaggeration of all those having centers below the dorsal cord (the patellar tendon and cremasteric are included). As to the reflexes having their center in the dorsal cord and which will be lost in the total destruction of the latter, the clinician recognizes only the abdominal.[151]

The abdominal wall retracts if the skin of the belly is lightly stroked (cutaneous reflex), or when this wall is percussed [tendon (?) reflex]. As studied by Rosenbach, Bodon, Parisot, Ostankoff, Dinkler, Pitres, etc., this reflex pertains to the region of the 9th (upper reflex), 10th, 11th, and 12th (middle and lower reflexes) pairs of intercostal nerves.

We will make the *differential diagnosis* after the following paragraph:

7. The Segmental Syndrome of the Dorsal Cord.

The study of medullary segmentalization has been made mostly from the clinical history of zonas of the trunk (Brissaud).[152]

Many authors[153] have already remarked clinically that thoracic zona is horizontal and often crosses the intercostal paths instead of being super-

---

[151]GANAULT. Th. citée. 1898, p. 102.

[152]BRISSAUD. Le zona du tronc et sa topogr. *Bull. méd.* 1896, p. 27 et 87.

[153]BRISSAUD cite: BAERENSPRUNG, BALMANNO SQUIRE, LEROUX, HEAD.

imposed upon them. Some have concluded from this that zona is not of nervous origin, others that the thoracic nerves have one course, "sensibly horizontal." Both opinions are impossible to support. Brissaud was the first to analyze and give a theory of the fact.

There are thoracic zonas of two nervous origins: Those of neuritic or ganglionic origin follow the course of the nerves, the others of medullary origin are horizontal. At the upper part of the thorax they are over the nerves, which are also horizontal. But in proportion as they descend the nerves become more and more oblique from above downward and from behind forward, and the zonas remain horizontal, crossing the nerves at more and more of an angle.

These horizontal zonas of the trunk represent very well the form of the zones of the distribution of the various segments of the dorsal cord. The neuralgias and the bands of anæsthesia, of hypæsthesia or hyperæsthesia, which present the same distribution, must be added here. Thus Achard[154] has described a band of dissociated anæsthesia, the topography of which was exactly that of an abdominal zona.

With these various symptoms the complete segmental syndrome of each segment of the dorsal cord is established.

8. The Root Syndrome of the Brachial Cord.

The *brachial cord*, which extends from the 4th cervical to the 2nd dorsal vertebra, corresponds to the origins of the 5th, 6th, 7th and 8th pairs of cervical and of the 1st pair of dorsal nerves.

Following is the table of the sensori-motor dis-

---

[154]ACHARD. Syringom. avec amyotr. du type Aran-Duchenne et anesth. dissociée en bande zostéroide sur le tronc. *Gaz. hebdom.* 1896, p. 361 et *Revue neurol.* 1896, p. 377.

The Diagnosis of

tribution borrowed from Testut, who has described it according to Thane (for motility) and Thornburn (for sensibility) :[155]

BRACHIAL CORD.

	Motor Functions.[156]	Sensibility.
5th. Cervical Pair. [Circumflex, sub-scapular, radial and musculo-cutaneous].	Longus colli, scalenus, angularis scapulæ, rhomboideus, serratus magnus, subclavius, sub and supra-spin ,teres min sub-scapularis,d biceps brachialis anticus.	Long band wbh corresponds to the external or radial side of the upper extremity; deltoid regi external surface of arm and forearm [outer part of then eminence].
6th. Cervical Pair. [Circumplex, sub-scapular, radial and musculo-cutaneous].	Longus coli, scalenus, serratus magnus, sub-capularis, deltoid, pectoralis major, biceps, brachialis anticus; pronator te palmas longus, supinato longus and brevis, et radial muscles, abductor, opponens and flexor brevi pollicis.	Middle portion anterior and posterior surfaces of thumb. } Thenar eminence and thumb.

[153] Voir aussi ALLEN STARR. *Brain* 1894, p. 481, *Revue neurol.* 1894, p. 570.

[156] Pour simplifier, je supprime, dans le tableau de Testut, les muscles à côté desquels il y a un point d'interrogation.

BRACHIAL CORD.

	Motor Functions.	Sensibility.
7th. *Cervical Pair.* [Radial].	Longus colli, post ior scal- enus and dor trice ficia exte exte posterio muscles.	Rest of the hand [ex- cept the hypothenar eminence] and the three middle fingers.
8th. *Cervical Pair.*	Longus colli, pectoralis major and minor, latissimus dorsi, triceps, anconei, flexors of the fingers, anterior ulnar muscles, pronator pollicis, interossei, flexor brevis and opposing muscles of little finger.	Long band along the internal or ulnar sur- face of upper extrem- ity: 1. Internal sur- face of arm, except small region near ax- illa, [2nd. dorsal]; 2. Internal surface fore- arm; 3. Hypothenar eminence and ring finger.
Part 1st. Dorsal. [Median and ulnar]	Pectoralis major and min- or; flexors of the fingers, ulnar muscles, pronator external and internal l, ser- ratus posticus superioris.	

The sensory root distribution of the various portions of the brachial cord may be easily derived by placing the subject on all fours, the upper extremity turned at an angle of 90 (in the primitive position), the thumb in advance. The sensibility is then seen to be distributed in three parallel bands (each occupying the entire length of the extremity), the posterior innervated by the 1st dorsal and 8th cervical, the middle by the 7th, and a part of the 6th cervical, the anterior by a part of the 6th and the 5th cervical.

Then, as for the lower extremity and the trunk, the zones of root distribution succeed each other in the same order and manner as the spinal pairs themselves.

For the motor distribution, it may be recalled that the lower pair (1st dorsal) correspond to the median and ulnar, the upper pairs (5th, 6th and 7th cervical) to the circumflex and radial. Marinesco has defined the position of these nuclei of origin in a recent work.[157] He shows notably that each nerve has a principal nucleus and some accessory nuclei, and that each nerve derives its origin from many medullary segments. "Thus the ulnar and median, the principal source of which is constituted by the eight cervical segment, still receive fibres from the 7th cervical and more from the 1st dorsal."

The *reflexes* which have their centers in this portion of the cord are the cutaneous and tendon reflexes of the upper extremity, the tendon reflex of the biceps having the highest medullary center (5th cervical).

In the same region is also the cilio-spinal center,

[157]MARINESCO. Contr. à l'étude des localisat. des noyaux moteurs, dans la moelle épin. *Revue neurol.* 1898. p. 463.

the presence of which is important in the semeiology of this region. The exciting oculo-pupillary fibres leave the cord by the 8th cervical and especially the 1st dorsal, and reach by the rami communicantes the inferior cervical sympathetic ganglion.

The experiments of Claude Bernard and Madame Déjerine-Klumpke (1885) show that when in an animal a section or laceration of the roots of the brachial plexus is made, pupillary phenomena are seen to follow whenever the rami communicantes of the first dorsal pair are injured, and only then.

There are also some confirmatory clinical facts. Raymond[158] cites those of Prevost, Pfeiffer, Heubner, Bruns, Monter, Muller, and particularly Sands and Seguin (1873). In the last case there was traumatic paralysis of the brachial plexus without oculo-pupillary symptoms. To stop the violent pains Seguin cut the lower roots of the plexus, and myosis appeared. Oppenheim was able to excite the first dorsal pairs in a man whose spine he had trephined. Excitation of the first dorsal pair alone determined a considerable mydriasis, which was maintained for some seconds.

Clinically, if the mydriasis indicates excitation of this cilio-spinal center (the lower part of the

[158]RAYMOND. Paral. radicul. du plexus brach. (7 déc. 1894) ; paral. radicul. sensit. du plex. brach. (22 mars 1895) ; un cas de paral. radicul. du plex. brach. dr. (24 avril 1896). *Leç. sur les mal. du syst. nerv.* 1896, t. I, p. 217 et 239; 1897, t. II, p. 379.

brachial cord), a destructive lesion of this same region of the cord would bring about myosis, narrow-
ing of the palpebral fissure and retraction of the
ocular globe.

From all this the description of the root syndrome of the brachial cord easily results: Pains,
paralysis, anæsthesias, amyotrophies, vaso motor
troubles and reflexes are distributed according to the
anatomo-physiological table which we give.

The syndrome may moreover be total or partial.
In the latter group there are many varieties, but two
principal types may be easily distinguished: the superior and inferior types.

In the superior type (Erb, Duchenne), the del-
toids, biceps, brachialis anticus and supinator
longus (Erb) are involved, and often also the supra
and sub-spinous muscles, the clavicular bundle of
the pectoralis major, the supinator brevis (Duchenne). It is a paralysis, partial or complete, of
the 5th and 6th cervical pairs.

In the inferior type (Klumpke) the muscles of
the median and ulnar are attacked: paralysis of the
1st dorsal pair.

The oculo-pupillary troubles serve only to establish the height of the lesion. They also permit one
to say whether the lesion is, in certain cases, in the
brachial plexus or in its roots. For the troubles do
not appear when the same nerves (8th cervical and
1st dorsal) are affected below the emergence of the
rami by the great sympathetic. This is then a potent
element in *differential diagnosis.*

9. The Segmental Syndrome of the Brachial Cord:

Brissaud derived the theory of medullary seg-·mentation from the segmental anæsthesias of the limbs in syringo-myelia and from the segmental zonas of the limbs.

For the *syringo-myelia*[159] he cites a case of Gilles de la Tourette and Zaguelmann (1889), and one of Parmentier (1890) of thermo-anæsthesia as gloves (long ones up to the elbow), then a case of Debove and one of Souques (1891) of dissociated anæsthesia as sleeves. He analyses the cases and shows that especially for the gloves, a part of three nerves, ulnar, median and radial, are necessary.

It is not the root distribution which we know to be in longitudinal bands parallel to the axis of the limb. It is the segmental distribution corresponding to the medullary segmentation.

For the *zonas*[160] it is also seen that the eruption presents a segmental distribution. Such are the cases of Head and Mankopf.

Brissaud next notices that it is the same in many cutaneous diseases. "Thus the case of *sebaceous ichthyosis* of Biefel has the transverse zones of syringo-myelia. On the arms in particular, two large cylindrical bracelets surround the middle portion of the humeral region." In a recent work on the relation of chronic *eczema* to anæsthesia of the skin, Stonkovenkoff and Nikolski have noted the existence of anæsthesias in symmetrical patches and per-

[159]BRISSAUD. Leç. sur les mal. nerv. 1895, p. 215.
[160]BRISSAUD. Sur la distribut. métamér. du zona des membr. *Presse méd.* 11 janvier 1896 et *Revue neurol* 1896, p. 710.

pendicular to the axis of the limbs in subjects hysterical or not hysterical.

Finally, *scleroderma*,[161] which may be distributed according to the track of a nerve or present a root topography, may also have a segmental distribution.

For the *motor nerves*, Joseph Collins[162] has already established that "the cell groups which give origin to the brachial plexus are three in number, and extend from the upper part of the 4th cervical pair to the lower part of the 1st dorsal. The cells of the upper part of this area supply the muscles of the shoulder and arm. The cells of the lower part supply the forearm and hand.

The nuclei of the flexors are external and at a lower level than those of the extensors. The cells which give rise to the nerves which innervate the extensors are situated nearer the median line than those which innervate the flexors.[163]

The same year from a case of muscular atrophy with autopsy, Graham M. Hammond[164] concludes also that one cellular group of the cord gives rise to the muscular nerves of the forearm and another group to those of the hand.

More recently, I myself[165] described a segmental

[161]BRISSAUD. Pathogénie du processus scléroderm. *Presse méd.* 1897, p. 285 (*Revue neurol.* 1897, p. 365) et DROUIN. Quelques cas de sclérod. local. à distrib. métamer. Th. Paris 1898.

[162]JOSEPH COLLINS. *The New-York med. Journ.* 1894, p. 40 et 98. *Revue neurol.* 1894, p. 105.

[163]On peut rapprocher ces données de celles que Marinesco a invoquées pour expliquer la fréquence de la "main de prédicateur" dans la syringomyélie.

[164]GROEME M. HAMMOND. *The New-York med. Journ.* 1894, p. 1. *Revue neurol.* 1894, p. 116.

[165]Congrès des neurolog. de Marseille, avril 1899. Voir aussi. nos leç. sur les Sympt. segmentaires de la moelle que le Dr Gibert va publier dans le *Nouveau Montpellier méd-*

tremor (hand and wrist) in a case of sclerosis in patches.

The segmental syndrome of the brachial cord is then thus characterized by sensory, trophic or motor symptoms having for their common character the involvement of a segment of the limb limited by a circular line perpendicular to the axis of the limb (line of amputation or of disarticulation).

10. The Syndrome of the Cervical Cord.

We include under the name *cervical cord* only the portion of the cord which corresponds to the first 3 vertical vertebræ and gives rise to the first 4 cervical pairs.

From the sensory point of view this medullary segment innervates the neck and the occipital region, above the region (already described) of the brachial and dorsal cords and limited by the region of the trigeminal. Its line of demarcation from the latter region[166] follows the inferior border of the lower jaw and the posterior border of its upright branch, passes in front of the ear and goes straight up to the top of the head to join the same line of the opposite side. In this sensory region of the cervical plexus the posterior part is innervated by the posterior branches (sub-occiptales) of the first 2 cervical nerves, and the anterior part by the anterior branches of the first 4 cervical nerves.

Here is the table (after Testut) of the motor distribution of the first 4 cervical pairs:[167]

---

ical et le fait d'amyotrophie en gant que vient de publier le Dr Crocq (*Journ. de neurol.*).

[166]Voir la fig. 485 de la p. 576 du t. II du Traité d'anat d. Testut, 3e édit. 1897.

[167]MAYET. Traité de diagn. méd. et de séméiol. 1898, t. I, 536.

		Anterior Branch.	*Posterior Branch.*
CERVICAL PLEXUS.	1st. *Cervical Pair.*	Lateral rectus, 2 anterior recti, genio-hyoid.	2 posterior recti, oblique and complexus.
	2nd. *Cervical Pair.*	Large anterior rectus, longus colli, sterno-cleido-mastoid.	Inf. oblique, complexus, splenius, sup. oblique.
	3rd. *Cervical Pair.*	Anterior rectus, longus colli, post scalenus, levator angularis scapulæ, trapezius.	Complexus, spinal muscles.
	4th. *Cervical Pair.*	Anterior rectus, longus colli, posterior scalenus, diaphragm, levator angularis scapulæ, trapezius.	Spinal muscles.

In résumé, for the sensibility, on putting the subject on four feet, the head hanging (the neck and occiput in front), the cervical plexus gives two bands, one anterior (the occipital region and the neck), the other posterior (the ear and anterior part of the neck).

The topography of the anæsthesias in lesions of this cord are naturally deduced from that. The pains are: 1st, painful torticollis, pains at the neck and at the occiput as a collar; 2nd, the length of the phrenic (from the base of the chest or in the shoulder with painful points at the level of the costal insertions of the diaphragm—7th to 10th rib—to the neck in front of the anterior scalenus, behind to the limit of the spinous processes of the 3rd and 4th cervical).

For motility, the cervical cord presides: 1st. over the various movements of the head on the trunk (flexion, rotation, extension); 2nd, over the movements of the diaphragm (phrenic). The symptoms of excitation or of paralysis of the first muscular group are easily foreseen: Convulsions (tics) or paralysis of rotation, of flexion or of extension of the head. • As to the diaphragm, Duchenne has established the symptoms of its paralysis.

In inspiration the epigastrium and hypochondriac regions are depressed instead of dilated, at the same time that the thorax increases in volume, and inversely during expiration. If there is simply a paresis, the phenomenon appears only in deep or agitated respirations; if the paralysis is unilateral, it appears only on one side. At the same time respiration is more frequent, especially on the least effort to walk or speak or on the least excitement. All the extraordinary muscles of inspiration become active then, the face is flushed and the patient suffocates.

The voice is feeble and the slightest utterance puts him out of breath. Expectoration is difficult or impossible. Defecation requires great efforts and is accomplished with much difficulty.

The *differential diagnosis* of this cervical syndrome should be made from the brachial or dorsal syndrome and from the bulbar syndrome.

The first will be made by the presence of symptoms which from the beginning pertain exclusively to the cervical cord and which we have described.

The second will be made by the absence of truly bulbar symptoms such as ocular paralyses, paralyses of the lips, tongue, deglutition, anæsthesias of the trigeminal, etc.

TABLE OF CONTENTS.

SOME BOOKS

ANATOMY—

BOWEN—

CHEEVER— ... Byron W. Cheever, A.M., M.D. ...
... in the University of Michigan ...
Clemes Smith Professor of ... ...
State School of Mines ... ...
edition. 12mo. $1.75.

FORD.—*The Cranial Nerves* ...
Professor of Anatomy and Physiology ...
Chart, 25 cents.

FORD.—*Classification of the Most Important* ...
Body. With Origin Insertion ...
of Each. By C. L. Ford, M.D. ...
Physiology in the University of Michigan ...

FRANCOIS. *1st Annotations* ...
and, Edited with Notes and Vocabulary ...
...

GRAY.—*Outline of Anatomy. A Guide to the Dissection of the Human Body, Based on Gray's Anatomy.* 54 pages. Leatherette, 60 cents.

HERDMAN-NAGLER.—*A Laboratory Manual of Electrotherapeutics.* By William James Herdman, Ph.B., M.D., Professor of Diseases of the Nervous System and Electrotherapeutics, University of Michigan, and Frank W. Nagler, B.S., Instructor in Electrotherapeutics, University of Michigan. Octavo. Cloth. 163 pages. 55 illustrations. $1.50.

HOWELL.—*Directions for Laboratory Work in Physiology for the Use of Medical Classes.* By W. H. Howell, Ph.D., M.D., Professor of Physiology and Histology. Pamphlet. 62 pages. 65 cents.

HUBER.—*Directions for Work in the Histological Laboratory.* By G. Carl Huber, M.D., Assistant Professor of Histology and Embryology, University of Michigan. Third edition, revised and enlarged. Octavo. 204 pages. Cloth, $1.50.

NOVY.—*Laboratory Work in Physiological Chemistry.* By Frederick G. Novy, Sc.D., M.D., Junior Professor of Hygiene and Physiological Chemistry, University of Michigan. Second edition, revised and enlarged. With frontispiece and 24 illustrations. Octavo. Cloth, $2.00.

NOVY.—*Laboratory Work in Bacteriology.* By Frederick G. Novy, Sc. D., M.D., Junior Professor of Hygiene and Physiological Chemistry, University of Michigan. Second edition, entirely rewritten and enlarged, 563 pages. Octavo. $3.00.

STRUMPELL.—*Short Guide for the Clinical Examination of Patients.* Compiled for the Practical Students of the Clinic, by Professor Dr. Adolf Strümpell, Director of the Medical Clinic in Erlangen. Translated by permission from the third German edition, by Jos. L. Abt. Cloth, 39 pages, 35 cents.

WARTHIN.—*Practical Pathology for Students and Physicians. A Manual of Laboratory and Post-Mortem Technic, Designed Especially for the Use of Junior and Senior Students in Pathology at the University of Michigan.* By Aldred Scott Warthin, Ph.D., M. D., Instructor in Pathology, University of Michigan. Octavo. 234 pages. Cloth, $1.50.

WARTHIN.—*A Blank Book for Autopsy Protocols.* By Aldred Scott Warthin, M.D., Ph.D., Assistant Professor of Pathology in the University of Michigan. Bound in Full Canvass, 50 cents.

CPSIA information can be obtained
at www.ICGtesting.com
Printed in the USA
BVHW051406051118
532204BV00004B/632/P